Advances in
MATHEMATICAL ECONOMICS

Aims and Scope. The project is to publish *Advances in Mathematical Economics* once a year under the auspices of the Research Center for Mathematical Economics. It is designed to bring together those mathematicians who are seriously interested in obtaining new challenging stimuli from economic theories and those economists who are seeking effective mathematical tools for their research.

The scope of *Advances in Mathematical Economics* includes, but is not limited to, the following fields:

- Economic theories in various fields based on rigorous mathematical reasoning.
- Mathematical methods (e.g., analysis, algebra, geometry, probability) motivated by economic theories.
- Mathematical results of potential relevance to economic theory.
- Historical study of mathematical economics.

Authors are asked to develop their original results as fully as possible and also to give a clear-cut expository overview of the problem under discussion. Consequently, we will also invite articles which might be considered too long for publication in journals.

S. Kusuoka, T. Maruyama (Eds.)

Advances in Mathematical Economics

The Workshop on Mathematical Economics 2009 Tokyo, Japan, November 2009
Revised Selected Papers

Volume 14

Springer

Shigeo Kusuoka
Professor
Graduate School of Mathematical Sciences
The University of Tokyo
3-8-1 Komaba, Meguro-ku
Tokyo 153-0041, Japan

Toru Maruyama
Professor
Department of Economics
Keio University
2-15-45 Mita, Minato-ku
Tokyo 108-8345, Japan

ISSN 1866-2226 ISSN 1866-2234
ISBN 978-4-431-54083-0 ISBN 978-4-431-53883-7 (eBook)
DOI 10.1007/978-4-431-53883-7
Springer Tokyo Dordrecht Heidelberg London New York

Printed on acid-free paper

Springer is part of Springer Science+Business Media (www.springer.com)

Preface

The present volume of *Advances in Mathematical Economics* is a collection of articles read at the Workshop on Mathematical Economics, which was held in Tokyo, November 13–15, 2009. The workshop was organized and sponsored by the Research Center for Mathematical Economics. On behalf of the organization committee, we would like to extend our deepest gratitude to Keio Gijuku Academic Development Funds and the Oak Society for their generous financial support, without which the workshop could not have been realized. It is, of course, with great pleasure that we express our warmest thanks to all participants of the workshop for their contribution to our project.

The Research Center for Mathematical Economics was founded in 1997. Thirteen years have already passed since then. To our delight, the Research Center has enjoyed frequent occasions to host conferences and meetings as well as to publish academic achievements of the researchers associated with the Research Center.

With deep regret, we recall some of our leading scientists who passed away during these thirteen years, including the late professor Gerard Debreu, the late professor Kiyosi Itô, and the late professor Leonid Hurwicz. It was sad for all of us that they were not present at the workshop in 2009.

We would like to dedicate this volume in their memory.

August 1, 2010

Shigeo Kusuoka
Toru Maruyama
Managing Editors
Advances in Mathematical Economics

Table of Contents

Research Articles

Adv. Math. Econ. 14, 1–16 (2011)

Advances in
MATHEMATICAL
ECONOMICS

How much can investors discount?

Takuji Arai and Takamasa Suzuki

Department of Economics, Keio University, 2-15-45 Mita, Minato-ku,
Tokyo 108-8345, Japan
(e-mail: arai@econ.keio.ac.jp)

Received: May 31, 2010
Revised: July 9, 2010

JEL classification: G10

Mathematics Subject Classification (2010): 91G99, 46N10, 91B30

Abstract. We suggest a new valuation method of contingent claims for complete markets. Since our new valuation is closely related to shortfall risk, our suggestion would be useful to study shortfall risk measures which are convex risk measures induced by shortfall risk. We firstly give a brief introduction of shortfall risk measures, and discuss a general form of the valuation. We shall then deal with diffusion type models which are complete market models with underlying assets described by diffusion processes. In particular, the valuation for American type claims is discussed.

Key words: Convex risk measure, shortfall, American type claims

1. Introduction

Throughout this paper, we consider a valuation method for contingent claims taking control of shortfall risk into account in the framework of complete market models. After giving a general form of the valuation, we shall deal with models whose underlying assets are described by diffusion processes, and obtain a result for American type claims.

Assuming our market is complete, we can believe that all claims have a fair price under the no-arbitrage condition. We consider a seller who intends to sell a claim. In spite of market completeness, we presume that the seller cannot sell it for its fair price for some reason. Then, a problem arises: How much can she discount it? If she sells it for a price less than its fair price, she would incur some shortfall risk. Hence, fixing the limit of her shortfall

S. Kusuoka, T. Maruyama (eds.), *Advances in Mathematical Economics Volume 14*,
DOI: 10.1007/978-4-431-53883-7_1,

risk which she can endure, she should control her cash flow not to exceed her limitation. In this setting, we shall obtain in this paper representation results of the least price which she can accept.

Firstly, we have to explain shortfall risk. We assume that the seller intends to sell a claim X, and her attitude toward risk is described by a loss function l. More precisely, l is a non-decreasing continuous convex function from $\mathbf{R}$ to $\mathbf{R}_+$ satisfying $l(x) = 0$ if $x \le 0$, and $l(x) > 0$ if $x > 0$. Let $\mathcal{U}$ be the set of all attainable claims with zero endowment. In many cases, the set $\mathcal{U}$ would be given by a set of stochastic integrations with respect to the underlying asset price process, or a set of random variables constructed by a stochastic integration minus a nonnegative random variable. Her shortfall risk, when she sells the claim X for a price $x \in \mathbf{R}$ and selects $U \in \mathcal{U}$ as her hedging strategy, is given by $E[l(-x - U + X)]$.

When her limit of shortfall risk is given by $\delta > 0$, the least price which she can accept would be described by

$$\inf\{x \in \mathbf{R} | \text{ there exists a } U \in \mathcal{U} \text{ such that } E[l(-x - U + X)] < \delta\}(=: \rho_l(-X)).$$

When we regard this as a functional ρ_l of $-X$, ρ_l is said to be a shortfall risk measure. Arai has investigated robust representations of shortfall risk measures in his papers [1] and [2] for general incomplete market cases. Roughly speaking, robust representations are given by using "sup" or "max" taken over a set of martingale measures. Thus, it would be so difficult to calculate concretely the values of shortfall risk measures for claims. On the other hand, in the complete market case, we do not have to take "sup" or "max", so that calculation would be comparatively easy. Thus, we shall restrict our market to be complete in this paper so as to obtain somewhat concrete results with respect to shortfall risk measures.

In Föllmer and Leukert [4], they considered some problems which are somewhat related to the one we shall treat in this paper. In particular, they discussed such problems in the framework of complete markets. The first is quantile hedging problem which maximizes the probability of a successful hedge, that is, $P(x + U \ge X)$ over $U \in \mathcal{U}$ under the constraint "$x \le$ a constant". In particular, they solved it by using the Neyman–Pearson lemma. In addition, they treated the problem minimizing the cost for a given probability of success, that is, minimizing $x \in \mathbf{R}$ such that there exists a $U \in \mathcal{U}$ satisfying $P(x + U \ge X) \ge c$, where c is a given constant in $(0, 1)$. Moreover, the Black–Scholes model was discussed in [4] as a common example of complete market models.

In Sect. 2, we review robust representations of shortfall risk measures. In particular, we shall introduce results in Arai [1] and [2]. Next, we deal with in Sect. 3 the complete market case. Moreover, we treat diffusion type models introduced in Karatzas and Kou [6] in Sect. 4. After describing the models, we shall discuss the Black–Scholes model as an example, and a valuation of American type claims.

2. Shortfall risk measures

In this section, we illustrate representation results on shortfall risk measures, which are introduced in Föllmer and Schied [5], and Arai [1] and [2].

Consider an incomplete financial market being composed of one riskless asset and d risky assets. The price process of the risky assets is given by an $\mathbf{R}^d$-valued RCLL special semimartingale S defined on a complete probability space $(\Omega, \mathcal{F}, P; \mathbf{F} = \{\mathcal{F}_t\}_{t\in[0,T]})$, where $T > 0$ is the maturity of our market, and $\mathbf{F}$ is a filtration satisfying the so-called usual condition, that is, $\mathbf{F}$ is right-continuous, $\mathcal{F}_T = \mathcal{F}$ and $\mathcal{F}_0$ contains all null sets of $\mathcal{F}$. Note that the process S is not assumed to be locally bounded. Let the interest rate be given by 0. Denote by $\mathcal{X}$ a suitable subset of L^0, the set of all random variables defined on $(\Omega, \mathcal{F}_T)$. Suppose that any contingent claim belongs to the set $\mathcal{X}$.

We presume a seller who intends to sell a claim $X \in \mathcal{X}$. We denote by l her loss function, and by $\delta > 0$ her limitation of shortfall risk. Henceforth, this limitation is called threshold. Let $\mathcal{U}$ be the set of all attainable claims with zero initial cost. Suppose that $\mathcal{U}$ is a convex set including 0. We shall regard any element $U \in \mathcal{U}$ as a hedging strategy. When X is priced for $x \in \mathbf{R}$ and a hedging strategy $U \in \mathcal{U}$ is selected, her shortfall and shortfall risk are defined by $(-x - U + X) \vee 0$ and $E[l(-x - U + X)]$, respectively. Then, a price x is called a good deal price of X for the seller, if there exists a $U \in \mathcal{U}$ such that $E[l(-x - U + X)] \leq \delta$. We can define good deal prices for a buyer by a similar way. The least good deal price for the seller gives the upper bound of a good deal bound induced by shortfall risk. See [1]. In this paper, we regard the least good deal price for the seller as a valuation of the claim. Defining a functional ρ_l on $\mathcal{X}$ as

$$\rho_l(X) := \inf\{x \in \mathbf{R} | \text{ there exists a } U \in \mathcal{U} \text{ such that } x + U + X \in \mathcal{A}^0\}, \quad (1)$$

where $\mathcal{A}^0 := \{Y \in \mathcal{X} | E[l(-Y)] \leq \delta\}$, the above least good deal price of the claim X is given by $\rho_l(-X)$. Föllmer and Schied [5] have proved that, roughly speaking, ρ_l defined by (1) becomes a convex risk measure in the framework of bounded claims and discrete time trading. Arai in [1] and [2] extended their result to the framework of Orlicz spaces and continuous time trading. In this section, we focus on introducing robust representation results of ρ_l on Orlicz spaces.

Now, we need to prepare terminologies and concepts on Orlicz spaces. A left-continuous non-decreasing convex non-trivial function $\Phi : \mathbf{R}_+ \to [0, \infty]$ with $\Phi(0) = 0$ is called an Orlicz function, where Φ is non-trivial if

$\Phi(x) > 0$ for some $x > 0$ and $\Phi(x) < \infty$ for some $x > 0$. When Φ is an $\mathbf{R}_+$-valued continuous, strictly increasing Orlicz function, we call it a strict Orlicz function in this paper. Note that, for any strict Orlicz function Φ, we have $\Phi(x) \in (0, \infty)$ for any $x > 0$ and $\lim_{x\to\infty} \Phi(x) = \infty$. Moreover, a strict Orlicz function Φ is differentiable a.e. and its left-derivative Φ' satisfies $\Phi(x) = \int_0^x \Phi'(u)du$. Note that Φ' is left-continuous, and may have at most countably many jumps. Define $I(y) := \inf\{x \in (0, \infty) | \Phi'(x) \geq y\}$, which is called the generalized left-continuous inverse of Φ'. We define $\Psi(y) := \int_0^y I(v)dv$ for $y \geq 0$, which is an Orlicz function and called the conjugate function of Φ. Any polynomial function starting at 0 whose minimal degree is equal to or greater than 1, and all coefficients are positive, is a strict Orlicz function. For example, cx^p for $c > 0$, $p \geq 1$, x^2+3x^5 and so forth. Moreover, $e^x - 1$, $e^x - x - 1$, $(x+1)\log(x+1) - x$ and $x - \log(x+1)$ are strict Orlicz functions. We define the following:

Definition 1. *For an Orlicz function Φ, we define two spaces of random variables:*
Orlicz space: $L^\Phi := \{X \in L^0 | E[\Phi(c|X|)] < \infty \text{ for some } c > 0\}$.
Orlicz heart: $M^\Phi := \{X \in L^0 | E[\Phi(c|X|)] < \infty \text{ for any } c > 0\}$.
In addition, we define two norms:
Luxemburg norm: $\|X\|_\Phi := \inf\left\{\lambda > 0 | E\left[\Phi\left(\left|\frac{X}{\lambda}\right|\right)\right] \leq 1\right\}$.
Orlicz norm: $\|X\|_\Phi^* := \sup\{E[XY] | \|Y\|_\Phi \leq 1\}$.

Remark that $M^\Phi \subset L^\Phi$ and both spaces L^Φ and M^Φ are linear. Moreover, if Φ is a strict Orlicz function, the norm dual of $(M^\Phi, \|\cdot\|_\Phi)$ is given by $(L^\Psi, \|\cdot\|_\Phi^*)$. In the case of the lower partial moments $\Phi(x) = x^p/p$ for $p > 1$, the Orlicz space L^Φ and the Orlicz heart M^Φ both are identical with L^p. In this case, the conjugate function is given by x^q/q, where $q = p/(p-1)$, and $M^\Psi = L^\Psi = L^q$. In general, if $\limsup_{x\to\infty} \frac{x\Phi'(x)}{\Phi(x)} < \infty$, then M^Φ is identical with L^Φ, for instance, $\Phi(x) = x - \log(x+1)$. Otherwise, M^Φ must be a subset of L^Φ, for example $\Phi(x) = e^x - 1$. Hereafter, a strict Orlicz function Φ is fixed to satisfy:

$$l(x) := \begin{cases} \Phi(x), & \text{if } x \geq 0, \\ 0, & \text{if } x < 0. \end{cases}$$

In other words, the loss function l is assumed to satisfy all conditions on strict Orlicz functions.

We can prove the following:

Proposition 1 (Proposition 3.3 of [1] and Theorem 2 of [2]). *Let $\mathcal{X}$ be given by L^Φ. Assuming that $\rho_l(0) > -\infty$, and the sequentially compactness of $\mathcal{U}$ in $\sigma(L^\Phi, L^\Psi)$, ρ_l is a $(-\infty, +\infty]$-valued convex risk measure on L^Φ, that is, ρ_l satisfies the following three conditions:*

(1) **Monotonicity:** $\rho_l(X) \geq \rho_l(Y)$ *for any* $X, Y \in L^\Phi$ *such that* $X \leq Y$.
(2) **Translation invariance:** $\rho_l(X+m) = \rho_l(X) - m$ *for* $X \in L^\Phi$ *and* $m \in \mathbf{R}$.
(3) **Convexity:** $\rho_l(\lambda X + (1-\lambda)Y) \leq \lambda\rho_l(X) + (1-\lambda)\rho_l(Y)$ *for any* $X, Y \in L^\Phi$ *and* $\lambda \in [0, 1]$.

When we take M^Φ instead of L^Φ, ρ_l is given by an $\mathbf{R}$-valued functional without the sequentially compactness of $\mathcal{U}$. Let $\mathcal{P}^\Psi$ be the set of all probability measures being absolutely continuous with respect to P and having L^Ψ-density with respect to P, that is, $\mathcal{P}^\Psi := \{Q \ll P | dQ/dP \in L^\Psi\}$. Corollary 1 of Biagini and Frittelli [3], together with Proposition 1, implies that ρ_l is represented as

$$\rho_l(X) = \sup_{Q \in \mathcal{P}^\Psi} \{E_Q[-X] - a_l(Q)\}, \tag{2}$$

where E_Q represents expectation under Q, and $a_l : \mathcal{P}^\Psi \to \mathbf{R}$ is the convex conjugate of ρ_l and is called the minimal penalty function. Remark that a_l is given by

$$a_l(Q) := \sup_{X \in L^\Phi} \{E_Q[-X] - \rho_l(X)\}. \tag{3}$$

From (2) and (3), we can prove representation results of ρ_l as follows:

Theorem 1 (Theorem 2 of [2]). *Under the same setting as Proposition 1, the shortfall risk measure ρ_l is represented as, for any $X \in L^\Phi$,*

$$\rho_l(X) = \sup_{Q \in \mathcal{P}^\Psi} \left\{ E_Q[-X] - \sup_{X^1 \in \mathcal{A}^1} E_Q[-X^1] \right.$$
$$\left. - \inf_{\lambda > 0} \frac{1}{\lambda} \left\{ \delta + E\left[\Psi\left(\lambda \frac{dQ}{dP} \right) \right] \right\} \right\}, \tag{4}$$

where

$$\mathcal{A}^1 := \{X^1 \in L^\Phi | \textit{ there exists a } U \in \mathcal{U} \textit{ such that } X^1 + U \geq 0 \; P\text{-a.s.}\}. \tag{5}$$

Remark 1. If we take M^Φ instead of L^Φ as the set $\mathcal{X}$, then we can change “sup” in (4) into “max”.

Corollary 1. *In Theorem 1, when $\mathcal{X} = M^\Phi$ and $\mathcal{U}$ is cone, we have*

$$\sup_{X^1 \in \mathcal{A}^1} E_Q[-X^1] = \begin{cases} 0, & \text{if } Q \in \mathcal{M}^\Psi, \\ \infty, & \text{if } Q \notin \mathcal{M}^\Psi, \end{cases}$$

where $\mathcal{M}^\Psi := \{Q \in \mathcal{P}^\Psi | E_Q[U] \leq 0$ *for any* $U \in \mathcal{U}\}$.

Proof. Note that $\sup_{X^1 \in \mathcal{A}^1} E_Q[-X^1] \geq 0$, since $0 \in \mathcal{U}$ and (5). Next, (5) implies that

$$\sup_{X^1 \in \mathcal{A}^1} E_Q[-X^1] \leq \sup_{U \in \mathcal{U}} E_Q[U].$$

If $Q \in \mathcal{M}^\Psi$, then $\sup_{U \in \mathcal{U}} E_Q[U] \leq 0$. Thus, $\sup_{X^1 \in \mathcal{A}^1} E_Q[-X^1] = 0$ for $Q \in \mathcal{M}^\Psi$. On the other hand, if $Q \notin \mathcal{M}^\Psi$, there exists a $U \in \mathcal{U}$ such that $E_Q[U] > 0$, which implies that $\sup_{U \in \mathcal{U}} E_Q[U] = +\infty$ by the cone property of $\mathcal{U}$. □

Corollary 2 (Corollary 4.3 of [1]). *Under the same assumptions as the previous corollary, for any $Q \in \mathcal{M}^\Psi$, if we find a $\widehat{\lambda}_Q > 0$ satisfying $\delta = E\left[\Phi\left(I\left(\widehat{\lambda}_Q \frac{dQ}{dP}\right)\right)\right]$, then we have*

$$a_l(Q) = E_Q\left[I\left(\widehat{\lambda}_Q \frac{dQ}{dP}\right)\right].$$

Recall that I is the generalized left-continuous inverse of the left-derivative Φ'. Note that we can find such a $\widehat{\lambda}_Q$ at least when I is continuous.

3. Complete market case

It would be difficult to calculate explicitly values of shortfall risk measures for a concrete model. If our market is complete, we do not have to take "sup" or "max" in (4). Thus, we treat in this section the complete market case as a simple one. More precisely, we presume a seller selling a claim H with loss function l and threshold δ. In the case where the seller cannot sell H for its fair price, she have to tolerate some shortfall risk. She then has to sell H for a price greater than or equal to $\rho_l(-H)$ to suppress her shortfall risk less than δ. Hence, we can regard $\rho_l(-H)$ as a valuation of H. In this section, assuming the market completeness, we shall calculate $\rho_l(-H)$ in the same setting as Corollary 1. We divide calculation into two steps. The first is the case where the function I, which is the generalized left-continuous inverse of the left-derivative Φ', satisfies the additional condition of Corollary 2. For more general cases, we shall adopt an approximating method under some mild conditions. Throughout this section, we suppose that $\mathcal{M}^\Psi = \{Q\}$ and $Q \sim P$, and $E_Q[U] = 0$ for any $U \in \mathcal{U}$.

In the first case, we have

$$\rho_l(-H) = E_Q[H] - a_l(Q) = E_Q[H] - \inf_{\lambda > 0} \frac{1}{\lambda}\{\delta + E[\Psi(\lambda\varphi)]\},$$

where $\varphi = dQ/dP$. If there exists a $\widehat{\lambda} > 0$ such that $E[\Phi(I(\widehat{\lambda}\varphi))] = \delta$, then Corollary 2 yields $\rho_l(-H) = E_Q[H] - E_Q[I(\widehat{\lambda}\varphi)]$. At least, such a $\widehat{\lambda}$ exists when I is continuous. Remark that $E_Q[H]$ is the fair price of H, and $E_Q[I(\widehat{\lambda}\varphi)]$ represents the penalty term, that is, the seller can discount H by $E_Q[I(\widehat{\lambda}\varphi)]$ off the fair price. Since our market is complete, we can find a replicating strategy for the claim $H - I(\widehat{\lambda}\varphi)$, which is denoted by $\hat{U}$. We have then

$$\begin{aligned}E[l(-\rho_l(-H) - \hat{U} + H)] &= E[l(E_Q[-H + I(\widehat{\lambda}\varphi)] - \hat{U} + H)]\\ &= E[l(I(\widehat{\lambda}\varphi))] = \delta,\end{aligned}$$

that is, $\hat{U}$ should be considered as the optimal strategy for the seller when she sells H for $\rho_l(-H)$. In summary, the valuation of H is equivalent to the fair price of $H - I(\widehat{\lambda}\varphi)$, and its optimal portfolio is given by the replicating portfolio $\hat{U}$. If the seller receives $-H+I(\widehat{\lambda}\varphi)$ at the maturity, then its shortfall becomes 0. However, since $I(\widehat{\lambda}\varphi)$ is, as it were, a virtual claim, she cannot receive it, which causes shortfall with size δ.

Next, we shall treat more general cases. That is, we consider the case where I may have jumps. Note that I has only at most countable jumps and never jump at 0. Let $j_0 := 0$ and $j_k, k \geq 1$ be the k-th jump point of I. Note that, if I has only $k(\geq 1)$ jumps, every $j_{k+l}(l \geq 1)$ becomes ∞. Denote $l_k := j_{k+1} - j_k$, for $k \geq 0$ such that $j_k < \infty$. In addition, denote $l'_k := \min\{l_k, 1\}$ for $k \geq 1$ and $J_n := \bigcup_{k=1}^{\infty} \left(j_k, j_k + \frac{l'_k}{n} \right)$. Now, we assume the following throughout this section:

Assumption 1. (i) $l_k > 0$ for any $k \geq 0$ such that $j_k < \infty$.
(ii) There exists a sufficient small $\varepsilon > 0$ such that we can take $\lambda = \lambda(\varepsilon) > 0$ to satisfy $E[\Phi(I((\lambda\varphi - \varepsilon) \vee 0))] \geq \delta$ and $E_Q[I(\lambda\varphi)] < \infty$.
(iii) The function I does not have a jump to ∞.

We assume Condition (iii) for simplicity. When $I(y)$ jumps to ∞, it is enough to consider as the domain of I only ys being less than the jump point of I to ∞ in the approximating method below. Thus, the above condition (iii) does not narrow models which we can treat in this section.

Let $\{I_n\}_{n\geq 1}$ be an increasing sequence of continuous functions which converges to I pointwise. We take each I_n for $n \geq 2$ to satisfy the following: $I(x) = I_n(x)$ on $\mathbf{R}_+ \backslash J_n$, $I(x) \geq I_n(x)$ on J_n and

$$\int_{j_k}^{j_k + \frac{l'_k}{n}} (I(x) - I_n(x))\, dx \leq \frac{1}{n^k} \text{ for any } k \geq 1.$$

Hence, we have

$$\int_0^\infty (I(x) - I_n(x))\, dx \le \sum_{k=1}^\infty \frac{1}{n^k} = \frac{1}{n-1}.$$

Defining $\Psi_n := \int_0^y I_n(z)dz$ for any $n \ge 2$, we have $0 \le \Psi(y) - \Psi_n(y) \le \frac{1}{n-1}$ and the sequence $\{\Psi_n\}$ is increasing.

Example 1. Let Φ be given by

$$\Phi(x) = \begin{cases} x, & \text{if } x \le 1, \\ x^2, & \text{if } x > 1. \end{cases}$$

We have then

$$\Phi'(x) = \begin{cases} 1, & \text{if } x \le 1, \\ 2x, & \text{if } x > 1, \end{cases} \quad \text{and } I(y) = \begin{cases} 0, & \text{if } y \le 1, \\ 1, & \text{if } 1 < y \le 2, \\ y/2, & \text{if } y > 2. \end{cases}$$

Thus, I has a jump at 1 from 0 to 1. Moreover, Ψ is given by

$$\Psi(y) = \begin{cases} 0, & \text{if } y \le 1, \\ y - 1, & \text{if } 1 < y \le 2, \\ y^2/4, & \text{if } y > 2. \end{cases}$$

Now, if we take a sequence $\{I_n\}$ as follows:

$$I_n(y) = \begin{cases} 0, & \text{if } y \le 1, \\ n(y-1), & \text{if } 1 < y \le 1 + \frac{1}{n}, \\ 1, & \text{if } 1 + \frac{1}{n} < y \le 2, \\ y/2, & \text{if } y > 2, \end{cases}$$

then all conditions on $\{I_n\}$ are satisfied and Ψ_n is given by

$$\Psi_n(y) = \begin{cases} 0, & \text{if } y \le 1, \\ \frac{n(y-1)^2}{2}, & \text{if } 1 < y \le 1 + \frac{1}{n}, \\ y - 1 - \frac{1}{2n}, & \text{if } 1 + \frac{1}{n} < y \le 2, \\ \frac{y^2}{4} - \frac{1}{2n}, & \text{if } y > 2. \end{cases}$$

Now, defining $\Phi_n(x) := \sup_{y \ge 0}\{xy - \Psi_n(y)\} \ge \Phi(x)$, the sequence $\{\Phi_n\}$ is decreasing, and $\Phi_n \to \Phi$ uniformly. Since each I_n is continuous, we can find a $\lambda_n > 0$ satisfying

$$E[\Phi_n(I_n(\lambda_n \varphi))] = \delta$$

and

$$\inf_{\lambda>0}\frac{1}{\lambda}\{\delta+E[\Psi_n(\lambda\varphi)]\}=E_Q[I_n(\lambda_n\varphi)].$$

We shall prove a key lemma as follows:

Lemma 1. *There exists a random variable A such that $\Phi_n(I_n(\lambda_n\varphi)) \to \Phi(A)$ in $\mathcal{L}^1$, taking a subsequence if necessary, that is, $E[\Phi(A)]=\delta$.*

Proof. We prove firstly the uniformly integrability of $\{\Phi_n(I_n(\lambda_n\varphi))\}_{n\geq 1}$. We fix a sufficient large n arbitrarily. Since $I_n(x)\geq I(x-1/n)$ by the definition of I_n, we have

$$\begin{aligned}\delta &= E[\Phi_n(I_n(\lambda_n\varphi))] \geq E[\Phi_n(I((\lambda_n\varphi-1/n)\vee 0))]\\ &\geq E[\Phi(I((\lambda_n\varphi-1/n)\vee 0))] \geq E[\Phi(I((\lambda_n\varphi-\varepsilon)\vee 0))].\end{aligned}$$

From Assumption 1, we have

$$E[\Phi(I((\lambda\varphi-\varepsilon)\vee 0))] \geq E[\Phi(I((\lambda_n\varphi-\varepsilon)\vee 0))].$$

Thus, $\lambda>0$ is greater than λ_n. We have then $\Phi_n(I_n(\lambda_n\varphi))\leq \lambda\varphi I(\lambda\varphi)+1$, which is in $\mathcal{L}^1$ by Assumption 1. Recall that $\Phi_n(x)\leq\Phi(x)+1$. As a result, $\{\Phi_n(I_n(\lambda_n\varphi))\}_{n\geq 1}$ is uniformly integrable.

Next, we prove that $\Phi_n(I_n(\lambda_n\varphi))\to\Phi(A)$ a.s. for some A. For any sufficient large n, we have $0<\lambda_n<\lambda$. Hence, λ_n has a subsequence converging to some $\lambda^*\in[0,\lambda]$. We denote such a subsequence by $\{\lambda_n\}$ again. Let $\varepsilon_0>0$ be fixed arbitrarily. For $n>m$, we have

$$\begin{aligned}&P(|\Phi_n(I_n(\lambda_n\varphi))-\Phi_m(I_m(\lambda_m\varphi))|>\varepsilon_0)\\ &\quad\leq P(|\Phi_n(I_n(\lambda_n\varphi))-\Phi_n(I_m(\lambda_m\varphi))|>\varepsilon_0/2)\\ &\qquad+P(|\Phi_n(I_m(\lambda_m\varphi))-\Phi_m(I_m(\lambda_m\varphi))|>\varepsilon_0/2)\\ &\quad=: K_1+K_2.\end{aligned}$$

Recall that $\Phi\leq\Phi_n\leq\Phi+\dfrac{1}{n-1}$. Thus, there exists an $n_0\in\mathbf{N}$ such that, for any $n>m\geq n_0$,

$$|\Phi_n(x)-\Phi_m(x)|=\Phi_m(x)-\Phi_n(x)\leq\varepsilon_0/2 \text{ for any } x\in\mathbf{R}_+.$$

That is, $K_2=0$ for any sufficient large n and m.

The set $\{x\in\mathbf{R}|P(\varphi=x)>0\}$, denoted by M, is at most countable. Letting $\varepsilon_1>0$ be fixed arbitrarily, we could select finitely many elements from M, which are denoted by $x_1,x_2,\ldots,x_N$, to satisfy

$$P\left(\bigcup_{x\in M\setminus\{x_1,x_2,\ldots,x_N\}}\{\varphi=x\}\right)<\frac{\varepsilon_1}{2}.$$

For any sufficient large n and m, we have

$$|\Phi_n(I_n(\lambda_n x_k)) - \Phi_n(I_m(\lambda_m x_k))| \leq \frac{\varepsilon_0}{2}$$

for $k = 1, \dots, N$, and

$$P(\{|\Phi_n(I_n(\lambda_n\varphi)) - \Phi_n(I_m(\lambda_m\varphi))| > \varepsilon_0/2\} \cap \{\varphi \in \mathbf{R}_+ \backslash M\}) < \frac{\varepsilon_1}{2}.$$

Hence, we have

$$\begin{aligned} &P(|\Phi_n(I_n(\lambda_n\varphi)) - \Phi_n(I_m(\lambda_m\varphi))| > \varepsilon_0/2) \\ &\quad < P\left(\bigcup_{x \in M\backslash\{x_1,x_2,\dots,x_N\}} \{\varphi = x\}\right) + P(\{|\Phi_n(I_n(\lambda_n\varphi)) - \Phi_n(I_m(\lambda_m\varphi))| \\ &\quad > \varepsilon_0/2\} \cap \{\varphi \in \mathbf{R}_+ \backslash M\}) < \frac{\varepsilon_1}{2} + \frac{\varepsilon_1}{2} = \varepsilon_1, \end{aligned}$$

from which $K_1 \to 0$ as $n, m \to \infty$ follows. As a result, we can conclude that

$$\lim_{n,m\to\infty} P(|\Phi_n(I_n(\lambda_n\varphi)) - \Phi_m(I_m(\lambda_m\varphi))| > \varepsilon_0) = 0,$$

that is, $\Phi_n(I_n(\lambda_n\varphi))$ converges to some $\Phi(A)$ in probability. Taking a subsequence if necessary, $\Phi_n(I_n(\lambda_n\varphi)) \to \Phi(A)$ a.s., namely, $E[\Phi(A)] = \delta$. □

We have then, for any sufficient large n,

$$a_l(Q) \geq \inf_{\lambda>0} \frac{1}{\lambda}\{\delta + E[\Psi_n(\lambda\varphi)]\} \geq a_l(Q) - \frac{2}{\lambda^*(n-1)},$$

since $\lambda_n \to \lambda^*$. Remark that λ^* is positive. Hence, by the definition of $\{\lambda_n\}$,

$$E_Q[I_n(\lambda_n\varphi)] = \inf_{\lambda>0} \frac{1}{\lambda}\{\delta + E[\Psi_n(\lambda\varphi)]\} \to a_l(Q)$$

as $n \to \infty$. Therefore, we can conclude as follows:

Theorem 2. *Under Assumption 1 and all conditions in this section, we have*

$$\rho_l(-H) = \lim_{n\to\infty} \{E_Q[H] - E_Q[I_n(\lambda_n\varphi)]\}.$$

Next, we calculate the optimal strategy for the seller when H sells for $\rho_l(-H)$. Firstly, we need to prepare the following lemma:

Lemma 2. *Taking a subsequence if necessary,* $I_n(\lambda_n\varphi)) \to A$ *in* $\mathcal{L}^1(Q)$*, that is,* $E_Q[A] = a_l(Q)$.

Proof. Since Φ^{-1} is a continuous function, Lemma 1 implies that

$$\Phi^{-1}(\Phi_n(I_n(\lambda_n\varphi))) \to A \text{ a.s.},$$

by taking a subsequence if necessary. For any $\varepsilon_1 > 0$, there exists a sufficient large number n_0 such that

$$P(\{|I_n(\lambda_n\varphi) - \Phi^{-1}(\Phi_n(I_n(\lambda_n\varphi)))| < \varepsilon_1 \text{ for any } n \geq n_0\}) = 1.$$

Hence, we have $I_n(\lambda_n\varphi) \to A$ a.s..

For any $n \geq 1$, we have $I_n(\lambda_n\varphi) \leq I_n(\lambda\varphi) \leq I(\lambda\varphi) \in \mathcal{L}^1(Q)$ by the definition of λ and Assumption 1. Thus, $\{I_n(\lambda_n\varphi)\}_{n\geq 1}$ is uniformly integrable in Q, which completes the proof of Lemma 2. □

Denote $\widetilde{\rho}_n(-H) := E_Q[H] - E_Q[I_n(\lambda_n\varphi)]$ and

$$l_n(x) := \begin{cases} \Phi_n(x), & \text{if } x \geq 0, \\ 0, & \text{if } x < 0. \end{cases}$$

Actually, $\widetilde{\rho}_n(-H)$ is the valuation of H for a seller with loss function l_n and threshold δ. Note that its optimal strategy is given by the replicating strategy $U_n \in \mathcal{U}$ for $H - I_n(\lambda_n\varphi)$. Since $\widetilde{\rho}_n(-H) \to \rho_l(-H)$ as $n \to \infty$, we could say that U_n approximates to the optimal strategy when H sells for $\rho_l(-H)$. Finally, we calculate the optimal strategy for a seller with l. Let U^A and U^H be the replicating strategies for A and H, respectively. Denoting $\hat{U} := U^H - U^A$, Lemmas 1 and 2 imply that

$$\begin{aligned} E[l(-\rho_l(-H) - \hat{U} + H)] &= E[l(-E_Q[H] + a_l(Q) - U^H + U^A + H)] \\ &= E[l(E_Q[A] + U^A)] = E[l(A)] = \delta. \end{aligned}$$

Consequently, $\hat{U}$ is the optimal strategy when H sells for $\rho_l(-H)$.

4. Diffusion type models

In this section, we consider diffusion type models constructed in Karatzas and Kou [6]. Firstly, we illustrate the diffusion type models.

A diffusion type model is a complete financial market model composed of one riskless asset and d risky assets. Assume that the interest rate is given by 0, that is, the price of the riskless asset is 1 at all times. Let $\{\mathbf{W}_t\}_{t\in[0,T]} = \{(W_t^1, \cdots, W_t^d)^*\}_{t\in[0,T]}$ be a d-dimensional Brownian motion, where $\mathbf{a}^*$ is the transposed vector of $\mathbf{a}$. Defining $\mathcal{F}_t^{\mathbf{W}} = \sigma(\mathbf{W}_s, 0 \leq s \leq t)$ for any $t \in [0, T]$, $\mathbf{F} = \{\mathcal{F}_t\}_{0\in[0,T]}$ is assumed to be given by the augmentation of

$\mathcal{F}^W$. For $i = 1, \cdots, d$, denoting by S^i the price process of the i-th risky asset, we suppose that S^i is given by a solution to the following SDE:

$$dS^i_t = S^i_t \left\{ b^i_t dt + \sum_{j=1}^d \sigma^{ij}_t dW^j_t \right\}, \quad t \in [0, T]$$
$$S^i_0 = s^i \in (0, \infty).$$

We suppose that the coefficient processes $\{\mathbf{b}_t\}_{t\in[0,T]} = \{(b^1_t, \ldots, b^d_t)^*\}_{t\in[0,T]}$ and $\{\boldsymbol{\sigma}_t\}_{t\in[0,T]} = \{(\sigma^{ij}_t)_{1\le i,j\le d}\}_{t\in[0,T]}$ are $\mathbf{F}$-progressively measurable and uniformly bounded in $(t, \omega) \in [0, T] \times \Omega$. In addition, we assume that $\boldsymbol{\sigma}_t$ is invertible and its inverse $\boldsymbol{\sigma}^{-1}_t$ is uniformly bounded in $(t, \omega) \in [0, T] \times \Omega$. We define an $\mathbf{R}^d$-valued process $\{\boldsymbol{\theta}_t\}_{t\in[0,T]}$, called the relative risk process, by $\boldsymbol{\theta}_t := \boldsymbol{\sigma}^{-1}_t \mathbf{b}_t$ for $t \in [0, T]$. The process $\boldsymbol{\theta}$ is bounded and $\mathbf{F}$-progressively measurable because of the assumptions on $\mathbf{b}$ and $\boldsymbol{\sigma}$. Under these assumptions together with Girsanov's theorem, the process Z defined by $Z_t := \exp\left(-\int_0^t \boldsymbol{\theta}^*_s d\mathbf{W}_s - \frac{1}{2}\int_0^t \|\boldsymbol{\theta}_s\|^2 ds\right)$, where $\|\cdot\|$ is the d-dimensional Euclidean norm, is a martingale and $\mathbf{W}^Q_t := \mathbf{W}_t + \int_0^t \boldsymbol{\theta}_s ds$ is an $\mathbf{R}^d$-valued Brownian motion under the probability measure Q defined by $Q(A) = E[Z_T 1_A]$, for any $A \in \mathcal{F}_T$. Note that Q is called the unique equivalent martingale measure.

A process $\{\boldsymbol{\pi}_t\}_{t\in[0,T]} = \{(\pi^1_t, \ldots, \pi^d_t)^*\}_{t\in[0,T]}$ is called a portfolio process, if it is an $\mathbf{F}$-progressively measurable process satisfying $\int_0^T \|\boldsymbol{\pi}_t\|^2 dt < \infty$ a.s.. Moreover, a cumulative consumption process $\{C_t\}_{t\in[0,T]}$ is defined as an increasing right continuous $\mathbf{R}$-valued $\mathbf{F}$-adapted process such that $C_0 = 0$, $C_T < \infty$ a.s.. Let $\mathcal{T}$ be the set of all stopping times on $[0, T]$. For any given portfolio/cumulative consumption process pair $(\boldsymbol{\pi}, C)$ and $x \in \mathbf{R}$, a solution $X := X^{x,\boldsymbol{\pi},C}$ to the linear stochastic equation

$$X_0 = x, \quad dX_t = \sum_{i=1}^d \frac{\pi^i_t}{S^i_t} dS^i_t - C_t,$$

is called the wealth process corresponding to initial capital x, portfolio process $\boldsymbol{\pi}$, and cumulative consumption process C. We call a portfolio/ consumption process pair $(\boldsymbol{\pi}, C)$ admissible with initial wealth x, if and only if there exists a nonnegative random variable Λ with $E_Q[\Lambda^p] < \infty$ for some $p > 1$ such that the wealth process $X := X^{x,\pi,C}$ satisfies

$$X^{x,\pi,C}_t \ge -\Lambda \text{ a.s., for any } t \in [0, T]. \tag{6}$$

We shall denote by $Adm(x)$ the class of such pairs. Although we might need a stronger integrability condition with respect to Λ depending on loss function

l, we describe the same one as [6] here for simplicity. For any $\tau \in \mathcal{T}$, we denote by $Adm(x, \tau)$ the class of portfolio/consumption process pairs (π, C) for which the stopped process $X^{x,\pi,C}_{\cdot\wedge\tau}$ satisfies the requirement (6).

4.1. Black–Scholes model

In this subsection, we consider the Black–Scholes model as a simple example of the diffusion type models introduced in the above. That is, we consider the case where $d = 1$, σ and b are constants, and $\sigma > 0$. In other words, the risky asset price process S is expressed by

$$S_t := s \exp\left\{\left(b - \frac{\sigma^2}{2}\right)t + \sigma W_t\right\}, \quad t \in [0, T],$$

where $s > 0$. We shall give a valuation formula of a claim H for a seller with exponential loss function and threshold δ, where we suppose that H satisfies an appropriate integrability condition. That is, her loss function l is given by

$$l(x) = \begin{cases} e^x - 1 & \text{if } x \geq 0, \\ 0 & \text{if } x < 0. \end{cases}$$

Then, its conjugate l^* and the inverse I of l' (the derivative of l) are denoted by

$$l^*(x) = xI(x) - l(I(x)) = \begin{cases} x\log x - x + 1, & \text{if } x \geq 1, \\ 0, & \text{if } 1 \geq x \geq 0, \end{cases}$$

and

$$I(x) = (l')^{-1}(x) = \begin{cases} \log x, & \text{if } x \geq 1, \\ 0, & \text{if } 1 \geq x > 0, \end{cases}$$

respectively. We calculate the value $\rho_l(-H)$. Note that Corollary 2 implies that $\rho_l(-H) = E_Q[H] - \inf_{\lambda>0} \frac{1}{\lambda}(\delta + E[l^*(\lambda Z_T)])$, where $Z_T = dQ/dP$. To calculate the second term in the RHS, we define a function f by $f(\lambda) = \frac{1}{\lambda}(\delta + E[l^*(\lambda Z_T)])$ for $\lambda > 0$. Then, there exists a unique positive number λ^* satisfying $E\left[(\lambda^* Z_T - 1)\, 1_{\{\lambda^* Z_T \geq 1\}}\right] = \delta$, and minimizing f with

$$f(\lambda^*) = \left(\log\lambda^* + \frac{1}{2}\left(\frac{b}{\sigma}\right)^2 T\right) h(\alpha) + \frac{b}{\sigma}\sqrt{\frac{T}{2\pi}} e^{-\frac{\alpha^2}{2}},$$

where $h(x) := \int_{-\infty}^{x} \frac{1}{\sqrt{2\pi}} e^{-\frac{y^2}{2}} dy$ and $\alpha := \frac{\sigma}{b\sqrt{T}} \log\lambda^* + \frac{b\sqrt{T}}{2\sigma}$ ($\alpha := \infty$ if $b = 0$). We can then conclude $\rho(-H) = E_Q[H] - f(\lambda^*)$.

4.2. American type claims

We shall try, in this subsection, to give a valuation method for American type claims in a diffusion type model. Karatzas and Kou in their paper [6] presented some basic results on the pricing problem for American type claims. Firstly, we review their results roughly.

Note that any American type claim is described as a process. Let $\{B_t\}_{t\in[0,T]}$ be an $\mathbf{R}$-valued process representing the payoff of an American type claim. Assume that the process B is $[0,\infty)$-valued, $\mathbf{F}$-adapted, having continuous paths, and satisfying

$$E^0\left[\sup_{t\in[0,T]} B_t^{1+\varepsilon}\right] < \infty \text{ for some } \varepsilon > 0.$$

Denoting the upper hedging price for B by h_{up}, and the lower hedging price by h_{low}, we can describe them as follows:

$$\begin{aligned} h_{up} &= \inf\{x \geq 0 | \exists(\hat{\pi},\hat{C}) \in Adm(x) \text{ s.t. } X_\tau^{x,\hat{\pi},\hat{C}} \geq B_\tau \text{ a.s.}, \forall\tau \in \mathcal{T}\}, \\ h_{low} &= \inf\{x \geq 0 | \exists\check{\tau} \in \mathcal{T}, (\check{\pi},\check{C}) \in Adm(-x,\check{\tau}) \text{ s.t. } X_{\check{\tau}}^{-x,\check{\pi},\check{C}} \\ &\qquad +B_{\check{\tau}} \geq 0 \text{ a.s.}\}. \end{aligned}$$

We define a function u on $[0, T]$ as

$$u(t) := \sup_{\tau\in\mathcal{T}_{t,T}} E_Q[B_\tau],$$

where $\mathcal{T}_{t,T}$ is the set of $[t, T]$-valued stopping times. They proved that $h_{up} = h_{low} = u(0)$. In other words, $u(0)$ is the fair price of B. Let $\hat{X}_t$, $\check{\tau}$ be as follows:

$$\begin{aligned} \hat{X}_t &:= \text{esssup}_{\tau\in\mathcal{T}_{t,T}} E_Q[B_\tau|\mathcal{F}_t], \text{ for } t \in [0,T], \\ \check{\tau} &:= \inf\{t \in [0,T) | \hat{X}_t = B_t\} \wedge T, \end{aligned}$$

respectively. It was proved that there exists a pair $(\hat{\pi},\hat{C}) \in Adm(u(0))$ satisfying the following:

$$\begin{aligned} X_t^{u(0),\hat{\pi},\hat{C}} &= \hat{X}_t \geq B_t, \text{ for } t \in [0,T], \\ X_t^{u(0),\hat{\pi},\hat{C}} &= -X_t^{-u(0),\check{\pi},0} > B_t, \text{ for } t \in [0,\check{\tau}), \\ X_{\check{\tau}}^{u(0),\hat{\pi},\hat{C}} &= -X_{\check{\tau}}^{-u(0),\check{\pi},0} = B_{\check{\tau}}, \hat{C}_{\check{\tau}} = 0, \end{aligned}$$

where $\check{\pi} := -\hat{\pi}$. Thus, they made it clear that $\hat{\pi}$, $\check{\tau}$, and $\hat{X}$ are the optimal hedging portfolio for a seller, the optimal exercise time for a buyer, and the price process of B, respectively. We can say that $u(0) = E_Q[B_{\check{\tau}}]$.

Now, we presume two investors. One is a seller of B with loss function l and threshold δ. Another is a buyer who intends to purchase B from the seller. We calculate a valuation of B from seller's view. Suppose that the seller intends to control her shortfall risk at the maturity. That is, the valuation should be given as the least price such that her shortfall risk at T is less than or equal to δ. Note that the seller cannot predict when the buyer exercises the claim B. Thus, she have to construct her hedging strategy no matter which stopping time is selected by the buyer, and can reconstruct her strategy after buyer's exercise. Hereafter, we assume that $\sup_{t\in[0,T]} B_t$ and the random variable Λ in (6) are in M^Φ, where Φ is the associated Orlicz function with l.

The valuation, denoted by $V(B)$, is given as follows. Now, we assume that any investor must take 0 as her consumption process. Let $Adm_0(x)$ be the set of portfolio processes $\boldsymbol{\pi}$ such that $(\boldsymbol{\pi}, 0) \in Adm(x)$. Next, we denote $X_t^{x,\boldsymbol{\pi}} := X_t^{x,\boldsymbol{\pi},0}$, $t \in [0,T]$ and $x \in \mathbf{R}$, and $X_{s,t}^{\boldsymbol{\pi}} := X_t^{0,\boldsymbol{\pi}} - X_s^{0,\boldsymbol{\pi}}$, $0 \le s < t \le T$. Moreover, for $\tau \in \mathcal{T}$ and $\boldsymbol{\pi} \in Adm_0(x)$, let $\widetilde{Adm}(\boldsymbol{\pi},\tau)$ be the set of portfolio processes $\tilde{\boldsymbol{\pi}}$ such that $E[l(-X_\tau^{x,\boldsymbol{\pi}} - X_{\tau,T}^{\tilde{\boldsymbol{\pi}}} + B_\tau)] < \infty$. We then define $V(B)$ as follows.

$$
\begin{aligned}
V(B) := \inf\{x \ge 0 | \exists \boldsymbol{\pi} \in Adm_0(x),\ \forall \tau \in \mathcal{T},\ \exists \tilde{\boldsymbol{\pi}} \in \widetilde{Adm}(\boldsymbol{\pi},\tau) \\
\text{s.t. } E[l(-X_\tau^{x,\boldsymbol{\pi}} - X_{\tau,T}^{\tilde{\boldsymbol{\pi}}} + B_\tau)] \le \delta\}. \qquad (7)
\end{aligned}
$$

Note that the value $X_\tau^{x,\boldsymbol{\pi}} + X_{\tau,T}^{\tilde{\boldsymbol{\pi}}} - B_\tau$ represents the final wealth of the seller with initial wealth x if the buyer exercises B at τ, the seller selects $\boldsymbol{\pi}$ as her hedging strategy at time 0 ($\boldsymbol{\pi}$ is independent of τ), and she changes her strategy into $\tilde{\boldsymbol{\pi}}$ at the moment τ that B is exercised. That is, if the seller sells B for a price greater than $V(B)$, then no matter which stopping time is selected by the buyer she can suppress her shortfall risk less than δ by selecting a suitable hedging portfolio.

Next, we try to obtain a representation of $V(B)$ by using the shortfall risk measure ρ_l. We consider the valuation of B when the seller postulates that the buyer exercises B at the optimal time $\check{\tau}$. Assume that the loss function l satisfies the additional condition of Corollary 2. Noting that $B_{\check{\tau}}$ is an $\mathbf{F}$-measurable random variable, we can consider $\rho_l(-B_{\check{\tau}})$. Thus, we rewrite the definition (1) as

$$
\rho_l(-B_{\check{\tau}}) = \inf\{x \ge 0 | \exists \boldsymbol{\pi} \in Adm_0(x) \text{ s.t. } E[l(-X_T^{x,\boldsymbol{\pi}} + B_{\check{\tau}})] \le \delta\}, \qquad (8)
$$

which represents the least price of B for the seller when she is certain that the buyer will exercise B at the optimal exercise time $\check{\tau}$. Corollary 2 implies that $\rho_l(-B_{\check{\tau}}) = E_Q[B_{\check{\tau}}] - E_Q[I(\hat{\lambda} Z_T)]$, where $Z_T = dQ/dP$, and $\hat{\lambda}$ is the unique positive constant satisfying $\delta = E[l(I(\hat{\lambda} Z_T))]$. Let $\boldsymbol{\pi}'$ be the replicating portfolio for $I(\hat{\lambda} Z_T)$, that is, $I(\hat{\lambda} Z_T) = X_T^{E_Q[I(\hat{\lambda} Z_T)],\boldsymbol{\pi}'}$ holds.

Now, we assume that $\hat{\pi} - \pi' \in Adm_0(\rho_l(-B_{\check{\tau}}) + \varepsilon)$ for any $\varepsilon > 0$, where $\hat{\pi}$ is the optimal strategy for B. Actually, we can say the following:

Proposition 2. $\rho_l(-B_{\check{\tau}}) = V(B)$.

Proof. By (7) and (8), it is clear that $\rho_l(-B_{\check{\tau}}) \leq V(B)$. We have $X_\tau^{\rho_l(-B_{\check{\tau}})+\varepsilon, \hat{\pi}-\pi'} + X_{\tau,T}^{-\pi'} - B_\tau > -I(\hat{\lambda} Z_T) \in \mathcal{A}^0$ for any $\varepsilon > 0$ and any stopping time $\tau \in \mathcal{T}$, since $X_\tau^{E_Q[B_{\check{\tau}}], \hat{\pi}} - B_\tau \geq 0$ for any $\tau \in \mathcal{T}$. Note that $-\pi'$ is in $\widetilde{Adm}(\hat{\pi} - \pi', \tau)$. Hence we have $\rho_l(-B_{\check{\tau}}) \geq V(B)$. □

This fact means that, even though the seller do not know when the buyer exercises B, the seller can select her hedging strategy as if the buyer necessarily exercised B at the optimal time $\check{\tau}$. Hence, for even American type claim B, its valuation induced by shortfall risk is represented by a shortfall risk measure. On the other hand, if the seller intends to control her shortfall risk at the moment that the buyer exercises, the problem would be more complicated. It remains to future research.

References

1. Arai, T.: Good deal bounds induced by shortfall risk. (2010, preprint)
2. Arai, T.: Convex risk measures on Orlicz spaces: inf-convolution and shortfall. Math. Financ. Econ. **3**, 73–88 (2010)
3. Biagini, S., Frittelli, M.: On the extension of the Namioka–Klee theorem and on the Fatou property for risk measures. In: Delbaen, F., Rasonyi, M., Stricker, C. (eds.) Optimality and risk: modern trends in mathematical finance. The Kabanov Festschrift. Springer, Berlin (2009)
4. Föllmer, H., Leukert, P.: Quantile hedging. Finance Stochast. **3**, 251–273 (1999)
5. Föllmer, H., Schied, A.: Convex measures of risk and trading constraints. Finance Stochast. **6**, 429–447 (2002)
6. Karatzas, I., Kou, S.: Hedging American contingent claims with constrained portfolios. Finance Stochast. **2**, 215–258 (1998)

Adv. Math. Econ. 14, 17–38 (2011)

Advances in
MATHEMATICAL
ECONOMICS

Variational analysis and mathematical economics 2: Nonsmooth regular economies

A.D. Ioffe

Department of Mathematics, Technion, Haifa 32000, Israel
(e-mail: alexander.ioffe38@gmail.com)

. . . In order for the analysis to be useful
it must provide information concerning
the way in which our equilibrium quantities
will change as a result of change of
parameters taken as an independent data

P. Samuelson,
"Foundations of Economic Analysis" 1947

Received: April 19, 2010
Revised: July 26, 2010

JEL classification: D11, C62

Mathematics Subject Classification (2010): 14P10, 49J52, 91B02, 91B42

Abstract. We introduce and study a quantitative concept of regularity (in the spirit of variational analysis) for exchange economies with set-valued demand correspondences and prove that in case the latter are semi-algebraic, every economy with the exception of a set of smaller dimension are regular. We also discuss the determinacy problem for such economies.

Key words: excess demand correspondence, equilibrium price, regular economy, coderivative, critical value of a set-valued map, semi-algebraic mapping

1. Introduction

This is the second of the block of two papers on applications of variational analysis to mathematical economics. In the first [20] we considered the model of welfare economics and applied methods of variational analysis to prove

S. Kusuoka, T. Maruyama (eds.), *Advances in Mathematical Economics Volume 14,*
DOI: 10.1007/978-4-431-53883-7_2,

an extremely general version of the second theorem of welfare economics which actually needed practically no restrictions on the choice of parameters of the economy such as preference relations and commodity spaces. Here we consider possible extensions of Debreu–Smale theorem on generic regularity of exchange economies [6]. The fundamental role of this theorem was forcefully emphasized in many subsequent publications (e.g. [8, 24]). Dierker [8] describes the two main questions studied by Debreu in [6] as follows:

(1) What does the set of equilibrium prices of an economy look like?
(2) How does the set of equilibrium prices of an economy E vary when the characteristic data of E vary?

and continues by explaining that to get the desirable answer "...one wants to establish the local uniqueness of each equilibrium price system and its continuous dependence on the characteristic data of the economy."

Local uniqueness of the equilibrium price system is usually referred to as *determinacy* and continuous dependence is related to what in the classical analysis would be called *regularity* of the mapping connecting equilibrium prices and economic parameters. The basic assumption behind Debreu's theory was that the individual demands of consumers are continuously differentiable. The mathematical model that appears in this case is a system of equations involving continuously differentiable functions in their left-hand sides and an equilibrium price is just a solution of the system. If the numbers of equations and unknowns coincide (which does happen in systems arising in Debreu's analysis) determinacy and regularity are equivalent properties by the inverse function theorem and the main conclusion that the economies with regular price systems fill an open set of full measure in the space of possible economies is based on the Sard theorem.

In the follow-up paper [7] (see also [24] for more details) Debreu specifies the requirements to the preference relation that guarantee that an individual demand is a continuously differentiable mapping. In the nutshell these requirements are the following: the preference relation is defined by a C^2 strictly quasi-concave and strictly monotone utility function without critical points and such that every indifference surface (level set of the function) have nonzero curvature at every point.

Weakening of the requirements leads to demands which may be even multivalued functions. And a natural question is whether there are some other suitable and meaningful classes of economies for which the same conclusion about typically good behavior of equilibrium price systems is possible. A positive answer comes from algebraic geometry which offers an extremely rich collection of non-pathological objects, so-called definable sets, functions and mappings, in particular semi-algebraic – the best known and most tractable. There are many arguments in favor of using these classes of objects in economic analysis (see e.g. [22] for a more detailed discussion). We just mention that the most popular types of utility functions do belong to them: Cobb–Douglas and CES utility functions, piecewise linear or exponential utilities – they all belongs to some classes of definable functions.

It seems appropriate to mention here that there is a growing interest to that class of sets and mapping in the optimization community which is actively working with non-differentiable and set-valued mappings for more than three decades. But to the credit of mathematical economists it should be said that the semi-algebraic and definable stuff attracted their attention much earlier: Blume and Zame [2] in 1993 extended Debreu's theorem to economies with preference relations associated with continuous quasiconcave definable utility functions. In one respect their statement is even stronger than that of Debreu: the set of "bad" economies is not just a set of measure zero but a definable set of smaller dimension – a reflexion of an important general property of definable sets.

The main technical tool used in [2] was the trivialization theorem saying roughly speaking that the graph of every definable set-valued mapping (e.g. from $\mathbb{R}^m$ into $\mathbb{R}^n$) is a union of finitely many homeomorphic images of graphs of constant set-valued mappings. The basic weakness of this techniques (which the authors explicitly indicate) is that it does not offer any mechanism of verification whether a specific point of the graph is regular or critical. Moreover, the very question of what is a regular or critical point or values of a non-differentiable or set-valued mappings, and so the entire regularity issue, remains outside the scope of the paper.

Meanwhile, in the regularity theory of variational analysis (often called metric regularity) this is the central question and infinitesimal characterization of regularity and even calculation of certain "measures of regularity" lie in the very heart of the theory. In this paper we offer a study of regularity and determinacy problems from the viewpoint of variational analysis.

We consider the same type of a problem as in the original paper by Debreu [6] in which the given data are individual demands (rather than utilities or preference relations) and we are not interested in how they have been obtained. In particular we do not consider the question of existence of equilibrium prices which mathematically is not actually connected with regularity. We assume that individual demands are set-valued mappings with definable graphs. In fact, we speak mainly about semi-algebraic sets and functions for the only reason that the definition of the latter is very simple and we give it in appropriate place, while the general definition of definable objects is much more involved and we have chosen not to quote it here. However all statements and proofs remain valid if we replace the word "semi-algebraic" by "definable".

The model is described in the next section. In Sect. 3 we briefly discuss the classical regularity concept and give a short proof of Debreu's theorem following the scheme of [8] rather than the original proof. Section 4 contains necessary information from local nonsmooth analysis. Section 5 is the central. In this section we introduce semi-algebraic sets and mappings and for them prove two theorems, one addressing regularity and the other local uniqueness of solutions (which for set-valued mappings are no longer equivalent) for semi-algebraic set-valued mappings.

The first theorem states that a regular value of a set-valued mapping $(x, p) \mapsto F(x, p)$ with semi-algebraic graph is also a regular value of partial mappings $x \mapsto F(x, p)$ for all p except maybe a semi-algebraic subset of the p-space whose dimension is strictly smaller than the dimension of the space. (A reader can easily recognize in this theorem a semi-infinite extension of a simplified version of the transversality theorem of Thom.) The second theorem says roughly speaking that for a semi-algebraic set-valued mapping from $I\!R^n$ into $I\!R^n$ and a regular value f of F solutions of the inclusion $y \in F(x)$ are locally unique if and only if the dimension of the graph of the mapping is n.

Finally, in Sect. 6 we return to our model and prove two theorems, one saying that for a typical (up to a closed set of smaller dimension) economy the set of equilibrium prices is either empty or displays Lipschitz dependence on variations of parameters of the economy, and the other giving a sufficient condition for typical determinacy of equilibrium prices in terms of dimensions of sections of the graph of the excess demand correspondence.

2. The model

We shall consider the simplest exchange economy with m agents and ℓ commodities. In the description of the model given below we use the following notation: $I\!R^\ell$ is the standard Euclidean ℓ-dimensional space with $x \cdot y$ being the inner product of x and y and $B(x, \alpha)$ is the ball of radius α around x, $I\!R^\ell_+$ is the nonnegative orthant, $I\!R^\ell_{++} = \text{int}\, I\!R^\ell_+$ is the collection of vectors with strictly positive components, $S^{\ell-1}$ is the unit sphere in $I\!R^\ell$ and $S_+ = S^{\ell-1} \bigcap I\!R^\ell_+$, $S_{++} = S^{\ell-1} \bigcap I\!R^\ell_{++}$. Finally, given sets X and Y, the symbol $X \rightrightarrows Y$ is used for set-valued mappings from X into Y.

The model of exchange economy with m agents and ℓ commodities is the collection of m triples $(X_i, \mathcal{D}_i, e_i)$, $i = 1, \ldots, m$, where

- $X_i \subset I\!R^\ell_+$ (we take it equal to the entire $I\!R^\ell_+$) is the *consumption set* of the i-th agent.
- $\mathcal{D}_i(p, w)$ is the *demand* of the i-th agent. This is generally a set-valued mapping which associates with each *price vector* p and the wealth w of the agent the possible choice of commodity vectors. It is assumed that $x \in \mathcal{D}_i(p, w)$ satisfies $p \cdot x = w$.
- $e_i \in I\!R^\ell_+$ is the *initial endowment* of the i-th agent.

We shall not be interested in the paper in how the demands of the agents have been determined. Usually it is done through preference relations or utility functions of the agents. If the preferences of the i-th agent are defined by the *preference correspondence* $\mathcal{P}_i(x) : X_i \rightrightarrows X_i$ ($u \in \mathcal{P}_i(x)$ means that u is strictly preferred to x), then $\mathcal{D}_i(p, w)$ is the collection of $x \in X_i$ satisfying the budget constraint $p \cdot x \leq w$ and such that no other affordable u is preferred to x, that is

$$x \neq u \in X_i, \quad p \cdot u \leq w \Rightarrow u \notin \mathcal{P}_i(x)$$

As the components of the price vector are nonnegative, the equality $p \cdot x = w$ for $x \in \mathcal{D}_i(p, w)$ is guaranteed if the preference relation is locally non-satiated (for any $u \in \mathcal{D}_i(p, w)$ the intersection of a neighborhood of u with $\mathbb{R}^l_{++}$ belongs to $\mathcal{P}(x)$).

When the preference is defined by means of a utility functions $u(\cdot)$ on $\mathbb{R}^\ell_+$ the demand is defined as the set of points of maximum of $u(x)$ on the budget set. But our arguments may be equally applied to the case of "generalized demands" in the spirit of Smale [29] when elements of the demand set are critical points of the maximization of the utility function over the budget set (that is satisfying the first order optimality condition, e.g. in terms of subdifferentials of non-smooth analysis – see e.g. [25, 28]).

The multivector $E = (e_1, \ldots, e_k) \in \mathbb{R}^{m\ell}_+$ is usually called *economy* and the set-valued mapping $(p, e) \to \mathcal{D}_i(p, p \cdot e)$ is the *demand correspondence* of the i-th agent. We observe that by definition $\mathcal{D}_i(\lambda p, \lambda p \cdot e_i) = \mathcal{D}_i(p, p \cdot e_i)$ for any $\lambda > 0$ (zero-degree homogeneity), so the natural domain of the demand correspondence is $S_+ \times \mathbb{R}_+$.

The set-valued mapping

$$Z(p, E) = \sum_{i=1}^{m} (\mathcal{D}_i(p, p \cdot e_i) - e_i)$$

from $S_+ \times (\mathbb{R}^\ell_+)^m$ into $\mathbb{R}^\ell$ is called the *excess demand* mapping. Since each $\mathcal{D}_i$ was supposed to satisfy $p \cdot \mathcal{D}_i(p, w) = w$ (that is $p \cdot x = w$ if $x \in \mathcal{D}_i(p, w)$) the *Walras law*: $p \cdot Z(p, E) = 0$ holds true.

Definition 1. *A vector $p \in \mathbb{R}^\ell_+$ is an equilibrium price (for the economy E) if $0 \in Z(p, E)$. We denote by $\mathbf{P}(E)$ the collection of equilibrium prices for E. The (generally set-valued) mapping $E \to \mathbf{P}(E)$ is called the equilibrium price correspondence of the model.*

The concept of regularity relates to the behavior of the equilibrium price correspondence. We postpone the formal introduction of this concept until we discuss it in the classical smooth situation in the next section and further in the final section the the general nonsmooth nonconvex model.

3. Classical regularity concept and Debreu's theorem

A continuously differentiable mapping $F : \mathbb{R}^m \to \mathbb{R}^n$ is *regular* at x if $F'(x)$, the derivative of F at x, maps $\mathbb{R}^m$ onto the whole $\mathbb{R}^n$. A vector $y \in \mathbb{R}^n$ is a *regular value* of F if either $y \neq F(x)$ for all x or F is regular at every $x \in F^{-1}(y)$. If y not a regular value of F it is called a *critical value* of F.

If y is a regular value of F belonging to the image of the mapping, then by the implicit function theorem $F^{-1}(y)$ is a smooth manifold of the same rank of smoothness as F (e.g. if $F \in C^k$, then $F^{-1}(y)$ is a C^k-manifold). The Sard theorem says that the collection of critical values of a C^k-mapping with $k > m - n \geq 0$ is a set of Lebesgue measure zero in $\mathbb{R}^n$. If in addition F is a proper mapping ($F^{-1}(Q)$ is a compact set if so is Q), then the set of critical values is a closed set of Lebesgue measure zero and $F^{-1}(y)$ is a compact set for any y.

Thus if $m = n$ and F is a proper C^1-mapping, the set $F^{-1}(y)$ may contain at most finitely many points for any regular value y of F and such y fill an open set of full measure in $\mathbb{R}^n$. Moreover, in this case y is a regular value of F if and only if $F^{-1}(y)$ is a finite set (or empty).

We shall be most interested in a situation when we have a mapping $F(x, w)$ from $\mathbb{R}^m \times \mathbb{R}^k$ with the second argument w viewed as a parameter. Suppose y is a regular value of F. Whether and when is y a regular value of a partial mapping $F_w(x) = F(x, w)$ for some w? This is actually a particular instance of the transversality problem. The following proposition is a key to answer.

Proposition 2. *Let F be a C^1-mapping from $\mathbb{R}^n \times \mathbb{R}^k$ into $\mathbb{R}^n$, and let $\overline{y}$ be a regular value of F. Consider the manifold $\Gamma = F^{-1}(\overline{y})$, and let φ stands for the restriction to Γ of the projection $(x, w) \to w$. If $\overline{w}$ is a regular value of φ, then $\overline{y}$ is a regular value of $F_{\overline{w}}$.*

Proof. Proofs of the propositions can be found in many publications (see e.g. [1, 16]). But we shall give it in view of its central role in our discussions. Let us denote by F_1' and F_2' the derivatives of F with respect to x and p respectively. Suppose $F(\overline{x}, \overline{w}) = \overline{y}$. As $\overline{y}$ is a regular value, for any $v \in \mathbb{R}^n$ there are $h \in \mathbb{R}^n$ and $q \in \mathbb{R}^k$ such that $F_1'(\overline{x}, \overline{w})h + F_2'(\overline{x}, \overline{w})q = v$.

On the other hand, as $\overline{y}$ is a regular value of F, the tangent space to Γ at $(\overline{x}, \overline{w})$ is $\{(e, s) \in \mathbb{R}^n \times \mathbb{R}^k : F_1'(\overline{x}, \overline{w})e + F_2'(\overline{x}, \overline{w})s = 0\}$. As φ is regular at $(\overline{x}, \overline{w})$, there is a $e \in \mathbb{R}^n$ such that $F_1'(\overline{x}, \overline{w})e + F_2'(\overline{x}, \overline{w})q = 0$. Thus $F_1'(\overline{x}, \overline{w})(h - e) = v$. This means (as v is an arbitrary element of $\mathbb{R}^n$) that the image of $F_1'(\overline{x}, \overline{w})$ is the whole of $\mathbb{R}^n$, that is $F_{\overline{w}}$ is regular at $\overline{x}$. □

As an immediate consequence we get the following simplified version of the transversality theorem of Thom.

Theorem 3. *Under the assumptions of Proposition 1, $\overline{y}$ is a regular value of F_w for all w outside of a set Lebesgue measure zero in $I\!R^k$.*

Proof. Indeed, $\dim \Gamma = k$, so the Sard theorem applies to φ. □

We also remark that in case when F_w is proper uniformly in w in a neighborhood of a certain $\overline{w}$, then the set $F_{\overline{w}}^{-1}(\overline{y})$ may contain at most finitely many points.

All said extends to the case of a mapping between two smooth manifolds.

These results are behind the seminal 1970 study by Debreu [6]. Debreu considered models with demand mappings $\mathcal{D}_i(p, w)$ assumed single-valued and continuously differentiable on $S_{++} \times I\!R_{++}$ and also satisfying the boundary *desirability assumption*:

(DA) $\|\mathcal{D}_i(p_n, w)\| \to \infty$ if one of the components of p_n goes to zero.

Thus Z is a C^1 mapping from $S_{++} \times I\!R^{m\ell}_{++}$ (which is a C^∞ manifold) into $I\!R^\ell$, so that the space $\mathcal{E}$ of all considered economies should be identified with $I\!R^{m\ell}_{++}$, and **(DA)** guarantees that the partial mapping $p \to Z(p, E)$ is proper locally uniformly with respect to $E \in \mathcal{E}$. There are several important points to be emphasized in connection with this model.

(a) If $e_i > 0$ for all i, then the desirability condition implies that $\mathbf{P}(E) \neq \emptyset$ (see [6]).

(b) Denote by $T(S_{++})$ the tangent bundle of S_{++} and by $T_p(S^{\ell-1})$ the tangent space of $S^{\ell-1}$ at p. As $p \cdot (\mathcal{D}_i(p, p \cdot e) - e) = 0$ the values of Z, for a given p, lie in the tangent space to $S^{\ell-1}$ at p. Thus, the graph of Z is a subset of the tangent bundle $T(S_{++})$. To make the situation formally compatible with the general setting considered in our brief above description of the regularity theory we just note that this tangent bundle $T(S_{++})$ is trivializable, that is there is a smooth mapping $\varphi : T(S_{++}) \to R^{\ell-1}$ such that $\varphi(p, \cdot)$ is a linear homeomorphism of $T_p(S_{++})$ and $I\!R^{\ell-1}$ for any $p \in S_{++}$ and the mapping $(p, h) \mapsto \varphi(p, h)$ is a diffeomorphism from $T(S_{++})$ onto $S_{++} \times I\!R^{\ell-1}$. Thus it is natural to call Z regular at a certain $(\bar{p}, \bar{E})$ if the composition $\psi(p, E)) = \varphi(p, Z(p, E))$ is regular at that point.

For instance, as all components of p are positive, the Walras law allows to take $\varphi(p, h)$ equal to the $(\ell - 1)$ dimensional vector of the first $\ell - 1$ components of h, so that $\varphi(p, Z(p, E))$ is just the vector of the first $\ell - 1$ commodities. In fact, there is no need to undertake any such action, because for any i the full partial derivative of $f_i(p, e) = \mathcal{D}_i(p, p \cdot e) - e$ with respect to e has rank $\ell - 1$. Indeed, setting $w = p \cdot e$, we have

$$\frac{\partial}{\partial e} f_i(p, e) = \frac{\partial}{\partial w} \mathcal{D}_i(p, w) p^T - I = A_i - I$$

where I is the identity matrix, and all vectors are viewed as columns, so that the transpose is a row vector. Every matrix A_i has rank one and $p^T(A_i - I) = 0$ and for any h orthogonal to p we have $(A_i - I)h = -h$.

These are well known arguments, of course (see e.g. [8]), but we need to keep them in mind for further discussions. Anyway, the conclusion is that *zero is a regular value of the excess demand correspondence.* We are now able to use Theorem 2 to get the principal Debreu's result of [6].

As has been agreed above, $\mathcal{E}$ is an open subset of $I\!R^{m\ell}$, hence a smooth manifold of the same dimension. As zero is a regular value of Z, the set $\Gamma = Z^{-1}(0)$ is also a C^1-manifold of dimension $m\ell$. It follows that Proposition 1 can be applied to the restriction to Γ of the projection $(p, E) \to E$. Applying Theorem 2, along with the mentioned existence theorem of [6], and taking into account that Z is proper in p we get

Theorem 4 (Debreu). *If all $\mathcal{D}_i$ are C^1-mappings satisfying (**DA**), then critical economies form a closed set of Lebesgue measure zero in $I\!R^{m\ell}$. Specifically, for every $E \in \mathcal{E}$ outside of a closed set of measure zero there are finitely many equilibrium prices and they smoothly depend on E.*

4. Regularity in variational analysis

Variational analysis studies nondifferentiable and even set-valued mappings. When we speak of a set-valued mapping $F : I\!R^m \rightrightarrows I\!R^n$, we admit that the empty set may be a possible value of the mapping as well as any other subset of the range space. The set $\mathrm{dom}\, F = \{x \in I\!R^m : F(x) \neq \emptyset\}$ is called the *domain* of F. In particular, the restriction of a single-valued mapping F to a subset $Q \subset I\!R^m$ can be viewed as a set valued mapping

$$F|_Q(x) = \begin{cases} F(x), & \text{if} \quad x \in Q; \\ \emptyset, & \text{otherwise.} \end{cases}$$

whose domain is Q.

The formal extension of the classical derivative based definition of regularity to these classes of mappings is therefore impossible. However there are equivalent descriptions of the phenomenon which can be applied also to set-valued mappings.

Proposition 5. *Let $F : I\!R^m \rightrightarrows I\!R^n$, and let $\overline{y} \in F(\overline{x})$. Then the following properties are equivalent:*

(i) *There are $\varepsilon > 0$ and $r > 0$ such that $B(y, rt) \subset F(B(x, t))$ if $y \in F(x)$, $\|x - \overline{x}\| < \varepsilon$, $\|y - \overline{y}\| < \varepsilon$, $0 \le t < \varepsilon$.*

(ii) *There are* $\delta > 0$ *and* $K > 0$ *such that* $d(x, F^{-1}(y)) \le Kd(y, F(x))$ *if* $\|x - \overline{x}\| < \delta,\ \|y - \overline{y}\| < \delta$.
(iii) *There are* $\gamma > 0$ *and* $N > 0$ *such that the function* $x \to d(y, F^{-1}(x))$ *satisfies on the ball* $B(\overline{x}, \gamma)$ *the Lipschitz condition with constant* N *if* $\|y - \overline{y}\| < \gamma$.
Moreover, if F *is single-valued and continuously differentiable, the three properties are equivalent to*
(iv) F *is regular at* $\overline{x}$.

The equivalence of (i)–(iii) is actually valid even for set-valued mappings between metric spaces. Certain forms of it were mentioned in some publications in the beginning of the 80s [9, 17], first proves were given in [4, 26], for a short prove see [18]. For a C^1 mapping the implication (iv) $\Rightarrow$ (i) is the famous Lusternik–Graves theorem actually contained in the main result of [15], the implication (iv) $\Rightarrow$ (ii) likewise follows from the "Lusternik theorem" of [21] and the main result of [27]. The opposite implications follow from Milyutin's perturbation theorem of [9] but for the first time they were explicitly explained in [10].

In fact, the equivalence between the three properties goes even further. Denote the upper bound of r in (i) by $\mathrm{sur}F(\overline{x}|\overline{y})$ (it is called the *modulus of surjection* of F at $(\overline{x}, \overline{y})$), the lower bound of K in (ii) by $\mathrm{reg}F(\overline{x}|\overline{y})$ (it is called the *modulus of metric regularity*) of F at $(\overline{x}, \overline{y})$) and the lower bound of N in (iii) by $\mathrm{lip}F^{-1}(\overline{y}|\overline{x})$ (it is called the *Lipschitz modulus* of F^{-1} at $(\overline{y}, \overline{x})$. The standard convention is to set $\mathrm{sur}F(\overline{x}|\overline{y}) = 0$ if (i) does not hold with any positive r and $\mathrm{reg}F(\overline{x}|\overline{y}) = \infty$ or $\mathrm{lip}F^{-1}(\overline{y}|\overline{x}) = \infty$ if K or N can be found for (ii) or (iii) to hold and agree that $0 \cdot \infty = 1$. Under these conventions the equalities

$$[\mathrm{sur}F(\overline{x}|\overline{y})]^{-1} = \mathrm{reg}F(\overline{x}|\overline{y}) = \mathrm{lip}F^{-1}(\overline{y}|\overline{x})$$

unconditionally hold for any F and any $(\overline{x}, \overline{y})$ in the graph of F. We also note that, as is immediate from the definition, the function $(x, y) \mapsto \mathrm{sur}F(x|y)$ is lower semi-continuous on Graph F and, accordingly, the modulus of metric regularity and Lipschitz modulus are upper semi-continuous.

In particular, if $A : \mathbb{R}^n \to \mathbb{R}^n$ is a linear homeomorphism, then the surjection modulus of A (is the same at every point of the domain space and) is equal to $\|A^{-1}\|^{-1}$.

Definition 6. Let $F : \mathbb{R}^m \rightrightarrows \mathbb{R}^n$ and $(\overline{x}, \overline{y}) \in$ Graph F. We say that F is *regular near* $(\overline{x}, \overline{y})$ or that $(\overline{x}, \overline{y})$ is a *regular point of* F if the three equivalent properties (i)–(iii) of Proposition 4 are satisfied. Otherwise, $(\overline{x}, \overline{y})$ is called a *critical point* of F. A $y \in \mathbb{R}^n$ is a *regular value* of F if (x, y) is a regular point whenever $y \in F(x)$.

Remark. It is important to keep in mind that according to the definition $\overline{y}$ is a regular point of F if $\overline{y} \notin F(x)$ for all x! On the contrary, any critical value belongs to the image of F.

In case of a single-valued mapping we do not mention y when speak about regular or critical points. Note that the definitions reduce to their classical counterparts if F is single-valued smooth. The only difference worth noting is that regular and critical points in the last definitions are points of the graph of the mapping rather than elements in the domain space. It easily follows from the definition that (for a set-valued mapping with closed graph) the set of critical points is closed. Thus if F is proper in the sense that $\|x_n\| \to \infty$ if $y_n \in F(x_n)$ and $\|y_n\| \to \infty$, then the set of critical values is also a closed set.

The following simple statement is extremely important for the understanding of the regularity phenomenon: *a set-valued mapping F is regular at* $(\overline{x}, \overline{y}) \in$ Graph F *if and only if the restriction to the graph of F of the projection* $(x, y) \to y$ *is regular at* $(\overline{x}, \overline{y})$ ([18], Proposition 1.3). The values of the regularity moduli of the last mapping depend on the choice of a specific equivalent norm in the product of the domain and range spaces but the statement is not connected with the choice.

An adequate mechanism for computing the moduli for mappings between finite dimensional spaces is provided by the subdifferential calculus. (This is already a well developed subject – see e.g. [11, 25, 28] for monographical accounts.) Given a function f on $I\!R^n$ finite at x, the *basic subdifferential* of f at x is

$$\hat{\partial} f(x) = \{y \in I\!R^n : \ y \cdot h \le f(x+h) - f(x) + o(\|h\|)\}$$

The inequality is supposed to be verified for all h of a small neighborhood of zero in $I\!R^n$. In this form the basic subdifferential is also known as the *Fréchet subdifferential*. Another form of the basic subdifferential, more suitable for computations, is known as *Dini–Hadamard* subdifferential:

$$\hat{\partial} f(x) = \{y \in I\!R^n : \ y \cdot h \le f^-(x; h, \ \forall\, h \in I\!R^n\},$$

where

$$f^-(x; h) = \liminf_{(t,u)\to(+0,h)} \frac{f(x+th) - f(x)}{t}$$

is *Dini–Hadamard lower directional derivative* of f at x. (Note that the Fréchet and Dini–Hadamard subdifferentials are no longer equal in infinite dimensional spaces.) The *limiting subdifferential* of f at x is

$$\partial f(x) = \limsup_{u \to_f x} \hat{\partial} f(u).$$

Here "$\limsup_{u \to_f x}$" stands for the graphical Painlévè-Kuratowski limit (all limit points of sequences (y_n) such that $y_n \in \hat{\partial} f(x_n)$ for some x_n such that $x_n \to x$ and $f(x_n) \to f(x)$). We observe that for a convex function the limiting subdifferential coincides with the subdifferential in the sense of convex analysis, whence the chosen notation.

The advantage of the limiting subdifferential over the basic subdifferential is that $\partial f(x) \neq \emptyset$ if f is Lipschitz near x. Another valuable quality of the limiting subdifferential (also shared by the basic subdifferential) is that it has convenient two ways relationship with the geometric concept of a normal cone. Recall, that given a set $S \subset I\!R^n$, the *indicator* of S is the function $i_S(x)$ equal to zero on S and ∞ outside of S. If $x \in S$, then $\partial i_S(x)$ is a cone which is called *limiting normal cone* to S at x and it is usually denoted $N(S, x)$. If on the other hand, we shall consider the distance function $d_S(\cdot)$ to S, then it turns out that $\partial d_S(x)$ coincides with the intersection of $N(S, x)$ with the unit ball. In the general case of a function f the equality

$$\partial f(x) = \{y \in I\!R^n \; : \; (y, -1) \in N(\text{epi } f, (x, f(x))\}$$

always holds. (Recall that epi $f = \{(x, \alpha) \in I\!R^n \times I\!R \; : \; \alpha \geq f(x)\}$ is the epigraph of f.)

Here are some elementary facts of the calculus of limiting normal cones:

$$N(Q \times P, (x, y)) = N(Q, x) \times N(P, y); \tag{1}$$

if $x \in Q \subset I\!R^m, \;\; y \in P \subset I\!R^n$;

$$N(Q \bigcap P, x) \subset N(Q, x) + N(P, x) \tag{2}$$

if (Q, P belong to the same space and) the following *qualification condition* holds:

$$u \in N(Q, x), \;\; v \in N(P, x), \;\; u + v = 0 \Rightarrow \; u = v = 0 \tag{3}$$

The last principal object of the subdifferential calculus, especially needed for characterizations of regularity is coderivative. Let $F \, : \, I\!R^m \rightrightarrows I\!R^n$ and $(x, y) \in$ Graph F. The *(limiting) coderivative* of F at (x, y) is the set-valued mapping

$$v \to D^* F(x, y)(v) = \{u \in I\!R^m \; : \; (u, -v) \in N(\text{Graph } F, (x, y)\}$$

In particular, if F is a linear operator, then the coderivative coincides with its adjoint.

The role of coderivatives in the regularity theory is revealed by the following theorem. Set Ker $D^* F(x, y) = \{v \, : \; 0 \in D^* F(x, y)\}$.

Theorem 7. *Let $F : \mathbb{R}^m \rightrightarrows \mathbb{R}^n$ be a set-valued mapping with locally closed graph, and let $y \in F(x)$. Then F is regular near (x, y) if and only if* Ker $D^*F(x, y) = \{0\}$. *Moreover,*

$$\text{sur}F(x|y) = \inf\{\|u\| : u \in D^*F(x, y)(v),\ \|v\| = 1\}.$$

For the specific case of F being the restriction of a linear operator to a convex cone we have

Corollary 8. *Let $A : \mathbb{R}^m \to \mathbb{R}^n$ be a linear operator, and let $K \subset \mathbb{R}^m$ be a closed convex cone. Let F stands for the restriction of A to K. Then*

$$\text{sur}F(0) = \min\{\|A^*y + u\| : \|y\| = 1,\ u \in K^\circ\},$$

where $K^\circ = \{u : u \cdot x \leq 0,\ \forall\, x \in K\}$.

5. Semi-algebraic and definable functions and mappings

The results presented in the previous section mainly relate to extensions to nonsmooth and set-valued mapping of the part of the classical regularity theory centered around the implicit function theorem. They are very general in that no real restriction is imposed on the behavior of mappings considered. In fact, as we have mentioned, they extend as far as to set-valued mappings between Banach spaces. There is no hope however that equally general extension is possible for the part of the classical theory centered around the Sard theorem. As in the classical case, we need to look for reasonable class of mappings for which an extension would be possible. By "reasonable" we mean "typical" or at least "suitable for various applications". A scale of convenient classes, so called *definable* objects, is offered by algebraic geometry. The simplest is the class of *semi-linear* sets and mappings: a semi-linear set is a finite union of convex polyhedra, either closed or open, a mapping is semi-linear if so is its graph. Note an extremely important role of polyhedral sets in optimization theory. Much richer is the class of *semi-algebraic* sets and mappings. A semi-algebraic set in $\mathbb{R}^n$ is a result of finite number of unions and intersections of sets of the form $\{x \in R^n : f(x) < 0\}$ or $\{x \in R^n : g(x) = 0\}$, where f and g are polynomial functions. As in the semi-linear case, a set-valued mapping is called semi-algebraic if so is its graph.

The list of the classes can be continued: we refer to [12] for the latest monographical account, in fact more classes of definable objects were discovered since then. We prefer not to give here the definition of definability which is not as straightforward as the definitions of semi-linear and semi-algebraic objects. So in what follows we shall speak about semi-algebraic sets and mappings but in any of the statements the change of the word "semi-algebraic" to "definable" is possible without reservations (except for some references, as certain results for semi-algebraic objects were known much before the very concept of definability was formed).

Semi-algebraic, mappings and functions have many remarkable properties (see e.g. [3, 5]). In particular, results of almost all major topological and analytic operations over such objects are also semi-algebraic. Specifically,

- The closure and interior of a semi-algebraic set is semi-algebraic.
- The derivative of a semi-algebraic function is semi-algebraic.
- The collection of semi-algebraic functions is stable w.r.t. algebraic operations (with a suitable convention concerning infinite values (e.g. $\infty - \infty = \infty$, $0 \cdot \infty = \infty$ etc.), pointwise maxima and minima.
- The image and preimage of a semi-algebraic set under semi-algebraic (set-valued) mapping is semi-algebraic.
- Composition of semi-algebraic (set-valued) mappings is a semi-algebraic mapping.
- The marginal function $\inf_x f(x, y)$ is semi-algebraic if so is f.

For our discussions the most important results about semi-algebraic sets and mappings is a deep stratification theorem proved in Łojasiewic in 1963 (see [23]). (To the general setting of definable sets and mappings the theorem was extended in [13].) Before stating the theorem we recall that a *Whitney stratification* of a set $Q \subset I\!\!R^n$ is a partition of Q into a finite number of smooth manifolds M_i (strata) which meet each other in a certain regular way, namely

(a) If $x_n \in M_j$ converge to an $x \in M_i$ $(i \neq j)$ and $h \in T_x M_i$ (the tangent space to M_i at x), then there are $h_n \in T_{x_n} M_j$ converging to h.
(b) If $(\mathrm{cl} M_j) \bigcap M_i \neq \emptyset$, then $M_i \subset (\mathrm{cl} M_j) \setminus M_j$.

If furthermore Q is partitioned into sets Q_j, then a stratification (M_i) of Q is said to be subordinate to the partition if for any i and j either $M_i \subset Q_j$ or $M_i \bigcap Q_j = \emptyset$.

Theorem 9 (Łojasiewicz stratification theorem). *Let $Q \subset I\!\!R^n$ be a semi-algebraic set, and let (Q_i) be a partition of Q into semi-algebraic sets. Then for any $k \in \mathbb{N}$ there exists a Whitney stratification of Q into finitely many semi-algebraic C^k-manifolds which is subordinate to (Q_i).*

(b) If F is a semi-algebraic (single-valued) mapping, then for any $k \in \mathbb{N}$ there is a C^k-Whitney stratification (M_i) of the domain of F such that the restriction of F to every M_i is k times continuously differentiable.

There may be of course many different stratifications. But the maximal dimension of the strata is the same for all stratifications; it is called the *dimension* of Q.

A fundamental consequence of the stratification theorem is that a Sard-type result is valid for semi-algebraic mappings, even in a stronger form than in the Sard theorem itself.

Theorem 10 ([19]). *Critical values of a semi-algebraic mapping $F : I\!R^m \rightrightarrows I\!R^n$ with locally closed graph form a semi-algebraic set of dimension not greater than $n-1$.*

(In fact, the requirement of a locally closed graph is not really necessary.)

In our sketch of the proof of the Debreu theorem we have used however not only the Sard theorem but also Proposition 2 which in turn is based on Proposition 1. But the direct extension of the proposition to semi-algebraic mappings is not possible.

Example 11. Consider the function

$$f(x,p) = |x| - |w|$$

viewed as a mapping from $I\!R^2$ into $I\!R$. This mapping is clearly semi-algebraic, even semi-linear, It is also an easy matter to verify that the mapping is regular at every point with the modulus of surjection identically equal to one (if we take the ℓ^∞ norm in $I\!R^2$). Furthermore

$$\Gamma = f^{-1}(0) = \{(x,w) : \ |x| = |w|\}$$

and the restriction to Γ of the projection $(x,w) \to w$ is also a regular mapping with the modulus of surjection equal one. However, the partial mapping $x \to f(x,0) = |x|$ is not regular at zero.

This means that we need to impose additional conditions to get an analog of Proposition 2 for semi-algebraic mappings which are not either single-valued or continuously differentiable.

Proposition 12. *Let $F : I\!R^m \times I\!R^k \rightrightarrows I\!R^n$ be a semi-algebraic set-valued mapping with locally closed graph. Let as above $\Gamma = F^{-1}(\overline{y})$ and $\hat{\Gamma} = \Gamma \times \{\overline{y}\}$. Suppose that there is an $(\overline{x},\overline{w}) \in \Gamma$ such that*

(a) F is regular at $((\overline{x},\overline{w}),\overline{y})$.
(b) The set-valued mapping $\Psi : I\!R^m \times I\!R^n \rightrightarrows I\!R^k$ defined by $\Psi(x,y) = \{w : \ y \in F(x,w)\}$ is regular at $((\overline{x},\overline{y}),\overline{w})$.
(c) There is a Whitney stratification (M_i) of Graph F *which is subordinate to the partition* Graph $F = \hat{\Gamma} \bigcup (\text{Graph } F \backslash \hat{\Gamma})$ *such that the restriction of the projection $(x,w) \to w$ to the set $S_i = \{(x,w) : (x,w,\overline{y}) \in M_i\}$, where M_i is the stratum containing $(\overline{x},\overline{w},\overline{y})$, is regular at $(\overline{x},\overline{w})$.*

Then $F_{\overline{w}}$ is regular at $(\overline{x},\overline{y})$.

Proof. Assume the contrary. Then by Theorem 6 there is a $v \neq 0$ such that $0 \in D^* F_{\overline{w}}(\overline{x}, \overline{y})(v)$. By definition this means that $(0, -v) \in N(\text{Graph } F_{\overline{w}}, (\overline{x}, \overline{y}))$. Consider the set $Q = \{(x, \overline{w}, y) \in I\!R^m \times I\!R^k \times I\!R^n : (x, y) \in \text{Graph } F_{\overline{w}}\}$. Up to the order of variables, this set is the product of Graph $F_{\overline{w}}$ and the singleton $\{\overline{w}\} \subset I\!R^k$. By (1) $N(Q, (\overline{x}, \overline{w}, \overline{y})) = \{(u, q, v) : (u, v) \in N(\text{Graph } F_{\overline{w}}(\overline{x}, \overline{y})),\ q \in I\!R^k\}$, simply is the product of $N(\text{Graph } F_{\overline{w}}(\overline{x}, \overline{y}))$ and $I\!R^k$ (again up to the order of variables).

On the other hand $Q = \text{Graph } F \bigcap \bar{P}$, where $\bar{P} = \{(x, w, y) : w = \overline{w}\}$. Clearly $N(\bar{P}, (\overline{x}, \overline{w}, \overline{y})) = \{0\} \times I\!R^k \times \{0\}$. We claim that

$$N(Q, (\overline{x}, \overline{w}, \overline{y})) \subset N(\text{Graph } F, (\overline{x}, \overline{w}, \overline{y})) + \{0\} \times I\!R^k \times \{0\}. \qquad (4)$$

To this end, following (2), (3), we have to show that $(0, q, 0) \in N(\text{Graph } F, (\overline{x}, \overline{w}, \overline{y}))$ only if $q = 0$. In order to see that this is the case, we first note that the graphs of F and Ψ coincide up to transposition of variables, y and w. Therefore

$$N(\text{Graph } \Psi, (\overline{x}, \overline{y}, \overline{w})) = \{(u, q, v) : (u, v, q) \in N(\text{Graph } F, (\overline{x}, \overline{w}, \overline{y}))\}.$$

But by the assumption Ψ is regular at $((\overline{x}, \overline{y}), \overline{w})$, so by Theorem 6 $(u, v) \neq 0$ if $(u, v, q) \in N(\text{Graph } \Psi, (\overline{x}, \overline{y}, \overline{w}))$ and $q \neq 0$. This proves the claim.

It follows that that there is a $q \in I\!R^k$ such that $(0, q, v) \in N(\text{Graph } F, (\overline{x}, \overline{w}, \overline{y}))$. As $v \neq 0$ and F is regular at $(\overline{x}, \overline{w}, \overline{y})$, we conclude, again applying Theorem 6, that $q \neq 0$.

Let now (M_i) be a Whitney stratification of Graph F satisfying (c), and let M_i be the stratum containing $(\overline{x}, \overline{w}, \overline{y})$. Then $(0, q, v) \in N(M_i, (\overline{x}, \overline{w}, \overline{y}))$ which is the same as $(0, q) \in N(S_i, (\overline{x}, \overline{w}))$ (as $M_i = \{(x, w, \overline{y}) : (x, w) \in S_i)\}$). But the latter cannot happen. Indeed, we can view S_i as the graph of a set-valued mapping from $I\!R^m$ into $I\!R^k$. This mapping is regular at $(\overline{x}, \overline{w})$ by (c) and, as we have seen, $q \neq 0$, so Theorem 6 excludes such a possibility. This completes the proof. □

We are now ready to prove a semi-algebraic version of Theorem 2.

Theorem 13. *Let $F : I\!R^m \times I\!R^k \rightrightarrows I\!R^n$ be a semi-algebraic set-valued mapping with locally closed graph, and let $\overline{y}$ be a regular value of F. Then for all $w \in I\!R^k$, with a possible exception of a semi-algebraic set of dimension strictly smaller than k, $\overline{y}$ is a regular value of the partial mapping $F_w : x \to F(x, w)$.*

Proof. Let $\Gamma \subset I\!R^n \times I\!R^k$ and $\hat{\Gamma} \subset I\!R^n \times I\!R^k \times I\!R^n$ be the same as in the proof of Proposition 12. Both are semi-algebraic set as so is Graph F. Let also the set-valued mapping $\Psi : I\!R^m \times I\!R^n \rightrightarrows I\!R^k$ be also as defined in Proposition 12(b). By Theorem 10 all $w \in I\!R^k$ with a possible exception of a semi-algebraic set of dimension smaller than k are regular values of Ψ.

Take an arbitrary Whitney stratification (M_i) of Graph F subordinate to the partition Graph $F = \hat{\Gamma} \bigcup (\text{Graph } F \backslash \hat{\Gamma})$. For any $M_i \subset \hat{\Gamma}$ set $S_i = \{(x, w) \in I\!R^n \times I\!R^k : (x, w, \overline{y}) \in M_i\}$ as in Proposition 12(c). Then S_i is a smooth semi-algebraic manifold, so applying again Theorem 10, we conclude that all $w \in I\!R^k$ with an exception of a semi-algebraic set of dimension smaller than k are regular values of the restriction to S_i of the projection $(x, w) \to w$.

The union of all exceptional sets (as there are finitely many of them) is also a semi-algebraic set of dimension smaller than k. For every $w \in I\!R^k$ outside this set and every $x \in I\!R^m$ such that $(x, w) \in \Gamma$ the conditions of Proposition 12 are satisfied if we set $\overline{x} = x$, $\overline{w} = w$ and application of Proposition 12 to all such w completes the proof. □

The last question we wish to discuss in this section is determinacy of F_w in case $n = m$, namely whether and under which condition $F_w^{-1}(\overline{y})$ is a finite set if $\overline{y}$ is a regular value of F_w. The answer to this question is both simpler and more complicated. It is simpler because a zero dimensional semi-algebraic set is automatically finite. On the other hand, it is clear that for a set-valued F the set $F_w^{-1}(\overline{y})$ can be very massive.

Theorem 14. *Let $F : I\!R^n \rightrightarrows I\!R^n$ be a semi-algebraic set-valued mapping. If* dim(Graph F) $\leq n$, *then for any regular value y of F the set $F^{-1}(y)$ is either empty or finite. Conversely, if the latter holds, then the intersection of* Graph F *with the product of the domain space and the set of regular values of F is a semi-algebraic set of dimension n. Moreover, in this case there is a natural N such that the cardinality of $F^{-1}(y)$ does not exceed N for any regular value of F.*

Proof. So let y be a regular value of F and $F^{-1}(y) \neq \emptyset$. We have to prove that the dimension of the set is zero. Assuming the contrary and using the facts that (a) a semi-algebraic set has finitely many semi-algebraic connected components and (b) a connected semi-algebraic set is semi-algebraically arcwise connected, we shall find a semi-algebraic curve $x(t)$, $t \in (0, a)$ such that $y \in F(x(t))$ for all t. As follows from the stratification theorem, we may assume without loss of generality that $x(\cdot)$ is a C^1-curve and that $\|\dot{x}(t)\| \equiv 1$. Set $z(t) = (x(t), y) \in I\!R^n \times I\!R^n$. Then $z(\cdot)$ is a C^1-curve in $I\!R^n \times I\!R^n$.

Let now (M_i) be a Whitney stratification of Graph F. We may assume that $z(\cdot)$ lies completely in one of the strata, say M_i. If $\dim M_i = n$, then the projection of the tangent space to M_i to the range space of F at every $z(t)$ must be surjective. (Indeed, in this case, the intersections of a small neighborhood of $z(t)$ with M_i and Graph F coincide. It follows that the restriction to M_i of the projection $(x, y) \to y$ is regular at every $z(t)$, whence the statement.) But as the dimensions of M_i and the range space coincide, we get that $v \neq 0$ for any nonzero $(h, v) \in T_{z(t)}M_i$. However $\dot{z}(t) = (\dot{x}(t), 0)$, – a contradiction.

Thus we have to conclude that $\dim M_i < n$. In this case, however, the restriction to M_i of the projection $(x, y) \to y$ is not regular at any point of M_i. On the other hand, every $z(t)$ is a regular point of the restriction of the projection to the graph of F. It follows that every $z(t)$ must belong to the boundary of some other stratum of dimension n.

So take a t, and let $z(t)$ belongs to the boundary of some M_j. Then there is a sequence $(z_n) \subset M_j$ converging to z. But then there are norm one vectors $w_n = (h_n, v_n) \in T_{z(t)}M_j$ converging to $(\dot{z}(t), 0)$. In other words, $\|v_n\| \to 0$. It follows that the projection of the unit ball in the tangent space $T_{z_n}M_j$ of M_j to the range space of F covers at most the ball of radius $\|v_n\|$. As the modulus of surjection is a lower semi-continuous function of its arguments, we conclude that $z(t)$ is a critical point of F. So we again get a contradiction which completes the proof of the first part of the theorem.

To prove the second statement, we may assume that all elements of the range space are regular values. Take a Whitney stratification (M_i) of Graph F and choose a manifold M_i with the largest dimension. Then any $(\overline{x}, \overline{y}) \in M_i$ has a neighborhood which does not meet any other element of the stratification. This means as above that $\overline{y}$ is a regular value of the restriction to M_i of the projection $(x, y) \to y$. Therefore the dimension of $\{x : (x, \overline{y}) \in M_i\}$ is $\dim M_i - n$. On the other hand the dimension of this set is zero as it is a subset of $F^{-1}(\overline{y})$. This proves the second statement. Finally the third follows from e.g. Lemma 2.13 (the uniform finiteness property) or Corollary 3.7 of [13]. □

6. Semi-algebraic model of an exchange economy

We now return to the model of exchange economy described in § 2. Throughout the section we suppose that the following *basic assumption* is satisfied:

- *The excess demand Z is a set-valued mapping from $S_+ \times I\!R_+^{m\ell}$ into $I\!R^\ell$ whose graph is a closed semi-algebraic set.*

We shall call an economy $\bar{E}$ *regular* if either $\mathbf{P}(\bar{E})$ is empty or there is a positive K such that $\mathcal{H}(\mathbf{P}(E), \mathbf{P}(E')) \le K\|E - E'\|$ for all $E,\ E'$ of a neighborhood of $\bar{E}$. Here $\mathcal{H}(Q, Q')$ stands for the Hausdorff distance between the sets Q and Q':

$$\mathcal{H}(Q, Q') = \inf\{r > 0 : \ Q \subset Q' + rB,\ Q' \subset Q + rB\}$$

(where B as usual is the unit ball). If $\bar{E}$ is not a regular economy, we call it *critical*.

In connection with this definition it is to be emphasized that we do not need the assumption that equilibrium prices exist for all $E \in I\!R_+^{m\ell}$. Moreover, nowhere in the subsequent discussions we even need the assumption that $Z(p,E) \neq \emptyset$ for every $(p,E) \in S_+ \times I\!R_+^{m\ell}$. It is sufficient to assume that the Walras law holds whenever $Z(p,E) \neq \emptyset$. Mathematically this underscores the fact that the existence and regularity problems are basically not connected. In particular, the assumption that the graph of the excess demand mapping is semi-algebraic which is crucial for the subsequent results does not seem to add anything for the existence of equilibrium prices.

Theorem 15. *Under the basic assumption, critical economies form a closed subset of $I\!R_+^{m\ell}$ whose dimension is strictly smaller than $m\ell$.*

Proof. We have Graph $\mathbf{P}(\cdot)$ is $Z^{-1}(0) = \{(p,E) \in S_+ \times I\!R_+^{m\ell} \ : \ 0 \in Z(p,E\}$. This is a closed semi-algebraic set. As the range space of $\mathbf{P}$ is compact, it follows that $\mathbf{P}$ is globally upper semicontinuous in the sense that for any $E \in$ dom $\mathbf{P}$ and any $\delta > 0$ there is an $\varepsilon > 0$ such that $\mathcal{H}(\mathbf{P}(\mathbf{E}'), \mathbf{P}(E)) < \delta$ if $\|E - E'\| < \varepsilon$.

Applying Theorem 10 to the inverse mapping $\mathbf{P}^{-1}$, we conclude that the collection of critical values of this mapping is a semi-algebraic set whose dimension is less than $m\ell$. Now let $\bar{E}$ be a regular value of $\mathbf{P}^{-1}$, and let $\bar{p} \in \mathbf{P}(\bar{E})$. By Proposition 5 the function $\varphi_E : E \mapsto d(p, \mathbf{P}(E))$ is Lipschitz for every p of a neighborhood of $\bar{p}$ and all E of a neighborhood of $\bar{e}$. The set $\mathbf{P}(\bar{E})$ is a closed subset of S_+, hence by the standard compactness arguments there are $\delta > 0$, $\varepsilon > 0$ and K such that for any p of the δ-neighborhood of $\mathbf{P}(\bar{E})$ φ is Lipschitz with constant K on $B(\bar{E}, \varepsilon)$. Taking a smaller ε, if needed, we can be sure that the sets $\mathbf{P}(E)$ corresponding to $E \in B(\bar{E}, \varepsilon)$ lie completely in the δ-neighborhood of $\mathbf{P}(\bar{E})$. But in this case the established Lipschitz property is equivalent to regularity of $\bar{E}$. □

As far as the determinacy question is concerned, in the general setting of this paper we only can refer to Theorem 14 which guarantees that at most finitely many equilibrium prices may exist for a certain economy E if zero is a regular value of the the mapping $p \mapsto Z(p,E)$ and the dimension of the graph of this mapping is $\ell - 1$. Verification of the dimension can only be based on some knowledge of the construction of the demand functions of all agents. Thus we may use this condition only as an assumption within the frameworks of this paper. As to regularity of the partial mapping $p \mapsto Z(p,E)$, we can refer to Proposition 12 which offers a possible mechanism for its verification.

But first of all we have to explain what we mean by a regular point and regular value of Z. As in the case of Debreu's theorem, $Z(p,E)$ is a subset of $(T_p S^{\ell-1}) \bigcap R_+^{\ell}$ for any p. The mapping $p \mapsto T_p S^{\ell-1}$ is not regular as a mapping into $I\!R^{\ell}$ at any point of the sphere, so again we consider a composition of Z with a trivialization map φ for the nonnegative sector of the tangent bundle. We can take for instance

$$\varphi(p,h)=\varphi(h)=\left(h^1-\frac{1}{\ell}\sum_{j=1}^{\ell}h^j,\ldots,h^\ell-\frac{1}{\ell}\sum_{j=1}^{\ell}h^j\right)$$

(here $h = (h_1,\ldots,h_\ell) \in I\!R^\ell$). This is a mapping onto the $(\ell-1)$-dimensional subspace $L_0 = \{h \in I\!R^\ell : h^1+\cdots+h^\ell = 0\}$. If $p \in S_+$, then the restriction of this mapping to T_pS is a linear isomorphism of T_pS and L_0 and this isomorphism is uniform on S_+ in the sense that its modulus of surjection is bounded away from zero by the same constants $\eta > 0$ for all $p \in S_+$. Let us agree to say that Z is regular at $(p,E) \in S_+ \times I\!R^{m\ell}_+$ if the composition mapping $Z_\varphi = \varphi \circ Z$ is regular at $(p,E,0)$ in the sense of Definition 6. We shall further say that zero is a regular value of Z if either $Z^{-1}(0)$ is empty or every $(p,E) \in Z^{-1}(0)$ is a regular point of Z.

Theorem 16. *We posit the basic assumption and in addition assume for all $E \in I\!R^{m\ell}_+$ with a possible exception of a semi-algebraic set of dimension smaller than $m\ell$ that the following two properties are satisfied:*

(a) $\dim\{(p,y) \in S_+ \times I\!R^\ell : y \in Z(p,E)\} \le \ell - 1$.
(b) If $p \in S_+$ and $p^j = 0$, then for any $i = 1,\ldots,\ell$ the j-th component of any element of $\mathcal{D}_i(p, p\cdot e_i)$ is equal to zero.

Then there is a natural N such that for all E with a possible exception of a semi-algebraic set of dimension strictly smaller than $m\ell$ the set $\mathbf{P}(E)$ can contain at most N points.

Proof. It is sufficient to consider only $E \in I\!R^{m\ell}_{++}$ for which the conditions hold (as the boundary of $I\!R^{m\ell}_{++}$ has dimension smaller than $m\ell$). Then, as immediately follows from (b), any element of $\mathbf{P}(E)$ must belong to S_{++}. As follows from Theorem 14, we should show that zero is the regular value of the partial mapping $p \to Z(p,E)$ for all $E \in I\!R^{m\ell}_{++}$ with possible exception of a semi-algebraic set of dimension smaller than $m\ell$. To prove the latter we have to show by Proposition 12 that zero is a regular value of the restriction of Z to $S_+ \times I\!R^{m\ell}_{++}$. Denote this restriction by $\check{Z}$. What we actually show is that under the assumptions $\check{Z}$ does not have critical values at all, that is that every $(p,E) \in \mathrm{dom}\,\check{Z}$ is a regular point of $\check{Z}$. This in turn will be proved if we show that for any $p \in S_+$ the partial mapping $E \mapsto Z(p,E)$ viewed as a mapping from $I\!R^{m\ell}_{++}$ into T_pS is regular at each point of its graph and this regularity is uniform in the following sense: for any $\bar{E} \in I\!R^{m\ell}_{++}$ there are $\varepsilon > 0$ and $r > 0$ such that the ball $B(\bar{E}, 2\varepsilon) \subset I\!R^{m\ell}$ of radius 2ε around $\bar{E}$ belongs to $I\!R^{m\ell}_{++}$ and for any $E \in B(\bar{E},\varepsilon)$, any $p \in S_+$ and any $x \in Z(p,E)$ we have

$$B(x,rt)\bigcap T_pS \subset Z(p,B(E,t)) \quad \text{if} \quad 0 \le t < \varepsilon. \tag{5}$$

Indeed in this case for any $y \in Z_\varphi(p, E)$ we have $B(y, \eta rt) \subset Z_\varphi(p, B(E, t))$ whenever $E \in B(\bar{E}, \varepsilon)$ and $0 \le t < \varepsilon$.

To prove this we shall show that (5) holds for any $p \in S_+$ and any $E \in I\!R^{m\ell}_{++}$ with ε depending only on the distance from E to the boundary of $I\!R^{m\ell}_+$. The latter can be properly represented by the distance from $e = \sum e_i$ to the boundary of $I\!R^+$ which is equal to the minimal component of the vector (recall that $e \in I\!R_{++}$ by the definition of $\mathcal{E}$).

So take a $E = (e_1, \dots, e_m) \in I\!R^{m\ell}_{++}$, set as above $e = \sum e_i$. We first consider the case of $p \in S_{++}$. By definition all components e^j, $j = 1, \dots, \ell$ are positive.

Take $t \le \min e^j$. Let finally $y \in Z(p, E)$, and let $h \in T_p S$, $\|h\| < t$. As $p \cdot h = 0$, we have $\mathcal{D}_i(p, p \cdot (e_i + \lambda_i h)) = \mathcal{D}_i(p, p \cdot e_i)$ if $\lambda_i \ge 0$. Take λ_i to satisfy $\sum \lambda_i = 1$ and let $E_h = (e_i + \lambda_i h)$. Then $Z(p, E_h) = Z(p, E) - h \supset y - h$. Thus, as $\|E_h - E\| \le \|h\| < t$, we see that $Z(p, B(E, t))$ contains the open t-ball around y. □

Passing to endowments with strictly positive components is fully justifiable as far as we wish to prove a generic fact. However, as Mas-Colell writes in [24], p. 336 "…it does not make sense in a reasonably rich model to allow the initial endowment vectors to be perturbed in any direction (implying that for a consumer to have a zero endowment of some commodity is a pathological situation!)." So it is reasonable to ask whether there are regular points of Z with endowments that may have zero component. The answer is positive and the proof can be modified to show that under the assumptions $(p, E) \in \text{dom}\, Z$ is a regular point of Z if $e = \sum e_i > 0$ and the total number of nonzero components of e_i is not smaller than $m + \ell - 1$.

We conclude with a brief comment concerning the conditions (a) and (b). The second of them is a weakened version of the desirability assumption (**DA**). The first condition (a) is more substantial. It was effectively proved in [2] in case when the preferences of agents were defined by continuous quasiconcave semi-algebraic utility functions. It seems however that quasiconcavity was basically used there to show that equilibrium prices do exist for every economy. We also mention in this connection a very recent result of [14] saying that the the dimension of the graph of the subdifferential mapping of a semi-algebraic function on $I\!R^n$ equals to n. But condition (a) may fail if the preference is not represented by a utility function. Consider, for instance the simple model in which the preference correspondence of every consumer is $\mathcal{P}(x) = x + I\!R^\ell_{++}$. Then $\mathcal{D}_i(p, w) = \{x \in I\!R^\ell : \ p \cdot x = w, \ x \ge 0\}$, in particular $e_i \in \mathcal{D}_i(p, p \cdot e_i)$ and therefore $\mathbf{P}(E) = S_+$ for all $E \in I\!R^{m\ell}_+$.

References

1. Aubin, J.P., Ekeland, I.: Applied Nonlinear Analysis. J. Wiley and Sons, New York (1984)
2. Blume, L.E., Zame W.R.: The algebraic geometry of competitive equilibrium. In: Neuefeind, W. (ed.) General Equilibrium and International Trade. In Memoraum Trout Rader, pp. 53–66. Springer, New York (1993)
3. Bochnak, J., Coste, M., Roy, M.F.: Real Algebraic Geometry, E.M.G. vol. 36. Springer, Berlin (1998)
4. Borwein, J.M., Zhuang, D.M.: Verifiable necessary and sufficient conditions for openness and regularity of set-valued and single-valued maps. J. Math. Anal. Appl. **134**, 441–459 (1988)
5. Coste, M.: An Introduction to O-Minimal Geometry. Inst. Rech. Math., Univ. de Rennes. http://name.math.univ-rennes1.fr/michel.coste/polyens/OMIN.pdf (1999)
6. Debreu, G.: Economies with a finite set of equilibria. Econometrica **38**, 387–392 (1970)
7. Debreu, G.: Smooth preferences. Econometrica **40**, 603–615 (1972)
8. Dierker, E.: Regular economies. In: Arrow, K.J., Intriligator, M.D. (eds.) Handbook of Mathematical Economics, vol. 3, pp. 795–830. North-Holland, Amsterdam (1982)
9. Dmitruk, A.V., Milyutin, A.A., Osmolovskii, N.P.: Ljusternik's theorem and the theory of extrema. Russ. Math. Surv. **35**(6), 11–51 (1980)
10. Dontchev, A.L.: The Graves theorem revisited. In: Mordukhovich, B.S., Sussmann, H.J. (eds.) Nonlinear Analysis and Geometric Methods in Deterministic Optimal Control, pp. 59–81. Springer, New York (1996)
11. Dontchev, A.L., Rockafellar, R.T.: Implicit Function and Solution Mapping. Springer, Berlin (2009)
12. van den Dries, L.: Tame Topology and O-Minimal Structures. Cambridge University Press, Cambridge (1998)
13. van den Dries, L., Miller, C.: Geometric categories and o-minimal structures. Duke Math. J. **84**, 497–540 (1996)
14. Drusvyatskiy, V., Lewis, A.S.: Semi-algebraic functions have small subdifferentials. (2010, preprint)
15. Graves, L.M.: Some mapping theorems. Duke Math. J. **17**, 111–114 (1950)
16. Guillemin, V., Pollack, A.: Differential Topology. Prentice Hall, Englewood Cliffs, NJ (1976)
17. Ioffe, A.D.: Non-smooth analysis: differential calculus of non-differentiable mappings. Trans. Am. Math. Soc. **255**, 1–55 (1981)

18. Ioffe, A.D.: Metric regularity and subdifferential calculus. Uspehi Mat. Nauk **55**(3), 103–162 (2000) (in Russian), English translation: Russ. Math. Surv. **55**(3), 501–558 (2000)
19. Ioffe, A.D.: Critical values of set-valued mappings with stratifiable graphs. Extensions of Sard and Smale-Sard theorems. Proc. Am. Math. Soc. **136**, 3111–3119 (2008)
20. Ioffe, A.D.: Variational analysis and mathematical economics 1. Subdifferential calculus and the second theorem of welfare economics. Adv. Math. Econ. **12**, 71–95 (2009)
21. Ioffe, A.D., Tihomirov, V.M.: Theory of Extremal Problems. Nauka, Moscow (1974) (in Russian); English translation: North Holland (1979)
22. Kubler, F., Schmedders, K.: Competitive equilibria in semi-algebraic economies. J. Econ. Theory **145**, 301–330 (2010)
23. Łojasiewicz, S.: Ensembles Semi-Analytiques. IHES Lecture Notes (1965)
24. Mas-Colell, A.: The Theory of General Economic Equilibrium. A Differentiable Approach. Cambridge University Press, Cambridge (1985)
25. Mordukhovich, B.S.: Variational Analysis and Generalized Differentiation. Springer, Berlin (2005)
26. Penot, J.-P.: Metric regularity, openness and Lipschitzean behaviour of multifunctions. Nonlinear Anal. **13**, 629–643 (1989)
27. Robinson, S.M.: Stability theory for system of inequalities. Part II: differentiable nonlinear systems. SIAM J. Numer. Anal. **13**, 497–513 (1976)
28. Rockafellar, R.T., Wets, R.J.B.: Variational Analysis. Springer, Berlin (1998)
29. Smale, S.: Global analysis and economics. J. Math. Econ. **1**, 1–14 (1974)

Adv. Math. Econ. 14, 39–67 (2011)

Advances in
MATHEMATICAL ECONOMICS

An overview of turnpike theory: towards the discounted deterministic case

M. Ali Khan[1] **and Adriana Piazza**[2,*]

[1] Department of Economics, The Johns Hopkins University, Baltimore, MD 21218, USA
(e-mail: akhan@jhu.edu)

[2] Departamento de Matemática, Universidad Técnica Federico Santa María, Avda. España 1680, Casilla 110-V, Valparaíso, Chile
(e-mail: adriana.piazza@usm.cl)

Received: July 22, 2010
Revised: July 30, 2010

JEL classification: C62, D90, Q23

Mathematics Subject Classification (2010): 91B62, 49J45, 49O2

Abstract. In the last 5 years, there has been extensive work on the existence and characterization of solutions of undiscounted optimal programs in simple discrete-time models of the 'choice of technique' in development planning and of lumber extraction in the economics of forestry. In this expository essay, we present a unified treatment of the characterization results in two of these models. Furthermore, with an eye towards extensions to the discounted setting, we present the general theory, both with or without "smoothness hypotheses" on the felicity function, and in continuous-time and discrete-time taking special care to distinguish asymptotic convergence of optimal programs from their classical turnpike properties. We show

* A preliminary version of this work was presented at the *Workshop on Mathematical Economics* held in Keio University, Tokyo during November 13–15, 2009. The authors are grateful to Professor Toru Maruyama for his invitation, and to him and Professors Alexander Ioffe, Alejandro Jofre, Boris Mordukhovich and Haru Takahashi for discussion and encouragement. Adriana Piazza gratefully acknowledges the financial support of Programa Basal PFB 03, *Centro de Modelamiento Matemático (CMM), Universidad de Chile* and that of FONDECYT under project 11090254. Ali Khan would like to acknowledge, in addition, the hospitality of the Departamento de Matemática, *Universidad Técnica Federico Santa María*, of the *CMM* and *Facultad de Economía y Negocios* at the *Universidad de Chile* during his visit in July 2010, when the final version of this work was completed.

S. Kusuoka, T. Maruyama (eds.), *Advances in Mathematical Economics Volume 14*,
DOI: 10.1007/978-4-431-53883-7_3,

how the general results do not translate directly to the particular toy models examined here, and thereby suggest open problems for a more universal theory.

Key words: Intertemporal resource allocation, discount factor, optimal programs, asymptotic convergence, classical turnpike theory, neighborhood turnpike theorem, development planning, forest management

> We wish to describe briefly what we shall mean by qualitative properties. The number ρ is called the *discount rate,* but we do not insist that ρ be less than one. As will be seen, the magnitude of ρ makes no difference to the analysis to follow. One need not even use the Kuhn–Tucker Theorem and can do the whole thing with elementary calculus.[1]
>
> Gale (1970)

1. Introduction

Turnpike theory, even without adjectival qualifications, is a somewhat diffused and misunderstood subject that deters entry to outsiders, hampers interdisciplinary dialogue between economists and mathematicians, and with its attendant terminological proliferation, renders the evaluation of marginal contribution of ongoing research exceedingly difficult to determine. It is not at all clear where the non-initiated reader, wanting to learn about the state of the subject and keep up with current work, ought to go. McKenzie's entry on *turnpike theory* in the 1987 *Palgrave* [13] is dropped in the 2008 version [12], and a search of the word *turnpike* in the latter turns up ten entries, none of them explicitly devoted to the theory and four on individual authors: Gale, Nikaido, Samuelson and Solow, with McKenzie and Radner conspicuously missing. Substantive entries which discuss the term are on capital theory [Becker], golden-rule [Phelps], (multisectoral) growth models [Brock–Dechert], neoclassical growth theory [Manuelli], and Pontryagin's principle of optimality [Zelkina], but except for the entry of Brock–Dechert, in each of these it turns up in a passing and cursory way. And, in another search engine, extensive references are furnished in Google Scholar to the *Townsend turnpike* in the field of monetary economics, an appropriation of the Samuelsonian term that takes it away to a total different subject matter. It is clear that the subject could do with an overview and the stables with some cleaning.

A point to begin is perhaps Bewley's 2007 text on *General Equilibrium, Overlapping Generations Models and Optimal Growth Theory* in which there are two sections devoted to the turnpike theorem in the final chapter, and the subject is introduced as follows:

---

[1] For the precise references to these sentences in [15], see Footnote 34 below.

> The term *turnpike theorem* arose early in work on optimal growth theory. We may visualize the turnpike in a model in which the initial and terminal capital stocks are held fixed. Suppose that the initial and terminal periods are -1 and T, respectively, so that $\underline{K}_{-1}$ and $\underline{K}_T$ are the initial and terminal capital stocks. Radner (1961) showed that if T was large, then the optimal path of capital approaches a unique optimal stationary level, stays close to it for a large fraction of the $T+2$ periods, and veers away toward the terminal stock only in the final periods. The image resembles a map of an interstate highway or turnpike with entrance and exit ramps.[2]

It is important to note that this part of the text is focussed on optimal finite-horizon programs, a focus that is relinquished in the subsequent sentences.

> In other work, researchers assume that the horizon is infinite, so that there is only an entrance ramp, and the theorem says that optimal paths asymptotically approach the unique stationary optimal path. The term *turnpike theorem* refers to any such assertion. Turnpike theorems give structure to optimal paths and focus attention on the stationary optimal one. Since convergence to the unique optimal stationary path is quite rapid in most fully specified examples, we are encouraged to think of modern economics as always near the stationary optimal state.

There are several lessons to be taken from this text. First, as already noted, the important distinction between finite-horizon and infinite-horizon resource allocation programs, and the corresponding inclusion of the latter with only an entrance ramp well within the rubric of turnpike theory. This inclusion is originally due to McKenzie,[3] and surely represents a substantial terminological over-reaching of the theory. Second, there is a focus on the uniqueness of the optimal stationary program, an assumption that is not necessary to turnpike theory, with or without an exit ramp, which at least according to McKenzie, can be viewed, rather broadly as the *bunching* or *clustering* of optimal trajectories around a certain benchmark set as the time horizon increases.[4] Third, the entire discussion is conducted in terms of

---

[2] See [4, pp. 542–543], also his Fig. 10.14 with time on the horizontal axis and capital stocks on the vertical.

[3] See [34] where McKenzie distinguishes three kinds of turnpikes: the *early, late* and *middle turnpikes.* In terms of this categorization, Bewley considers only the last two kinds of turnpikes. It is also worth pointing out that in [39], McKenzie uses a different classification scheme based on the *Samuelson* and *Ramsey turnpike.* It may be worth stating that in this essay, we are using the word as an adjective, as in "turnpike theory" or in "turnpike theorem", rather than as a noun.

[4] In this connection, see [25] for detailed references to the McKenzie's *oeuvre.*

discrete-time resource allocation problems, and the asymptotic implementation of the continuous-time results as being derived from their discrete-time counterparts is not raised. This must surely remain an important desideratum for the theory, and especially when the subject moves to its *raison d'etre* in computation. Finally, a complete silence as regards the discount factor being a crucial determinant of the shape of the theory; the independence of the dynamics on it.

As regards the discount factor, Bewley presents a result in the discounted setting on the global asymptotic stability of the solution to a one-sector infinite-horizon growth model due to Ramsey (1928), Cass (1965), Koopmans (1965) and Samuelson (1965).[5] He writes:

> A disappointing feature of growth theory is that this turnpike theorem does not generalize fully to models with many commodities or with uncertainty. The turnpike theorem with a positive discount rate, r, seems to require special assumptions in these interesting settings, and the generalizations are true only when r is sufficiently small. The turnpike theorems with $r = 0$ do generalize to models with many commodities and uncertainty. For this reason they probably should be thought of as the true turnpike theorems.[6]

And again.

> Although the theory of optimal allocations and programs becomes more difficult when utility is not discounted, the theory does become more robust. For instance, the turnpike theorem does not apply fully when there is more than one commodity and future utility is discounted, but it does apply when there are multiple commodities and there is no discounting.[7]

In his 1987 *Palgrave* entry, McKenzie had already noted that the "spirit of the original turnpike theorem is not well preserved in the aggregative model since the emphasis in the original theorem lies on the relative composition of the capital stock."

This emphasis on the spirit of the original theorem is precisely the point of departure for this essay. The original Samuelsonian motivation of turnpike theory was to cope with the indeterminacy of the terminal capital stock in finite-horizon resource allocation problems by arguing for its irrelevance

5 Also referred to as the "aggregate model", or in the macroeconomic literature, as the Ramsey–Cass–Koopmans (RCK) model.

6 See [4, p. 544] where Bewley refers to his Theorems 10.79 and 10.80 in this connection.

7 See [4, p. 551] where the fact that the undiscounted "theory with multiple commodities" is not presented is footnoted. Also see a repetition of this statement on page 577.

in the setting of a "large enough" finite-horizon. The two turnpike theorems that Bewley presents for the undiscounted setting appropriate the term solely for the global asymptotic stability of solutions to infinite-horizon programming problems, and concern the "overlapping generations model" and the one-sector aggregate growth model.[8] To be sure, the optimal stationary program, being a solution to a static optimization problem, is a possible benchmark for either the finite- or the infinite-horizon problem, and the emphasis on an infinite-horizon problem is a complementary approach to the difficulty mentioned above of deciding on the 'optimal value of the terminal capital stock'.[9] Both approaches draw the relevance of the von Neumann maximal growth path, or the stationary state in the writings of the classical economists before him, to optimal solutions to resource allocation problems.[10] But there is an important analytical difference. In the first case, this relevance is established by turnpike theorems for solutions to finite-horizon problems, and in the second, by asymptotic stability theorems for solutions to infinite-horizon problems. The first involves explicit quantification of the unknowns, and is logically antecedent to the qualitative treatment involved in the second approach.[11] As we shall see in the sequel, this appropriation is very much in tune with the current research literature. This may be dismissed as a semantic issue hardly of any substantive import.[12] However, one of the reasons for striving for terminological clarification is also for the sake of communication: simply to be clear to oneself and to one's readers the precise subject matter that is being discussed. Furthermore, a rich terminology for asymptotic stability is already available in the classical theory of difference and differential equations.[13] Anyhow, what is especially interesting about Bewley's text is not its discussion of turnpike theory in the large but in its integration of static and dynamic general equilibrium theory, a 2007 follow-through perhaps of the

---

[8] These are the theorems referred to in Footnote 6. Subsequent to their presentation, he re-emphasizes the meaning of the phrase "true turnpike theorems"; see [4, p. 577, paragraph 4].

[9] See, for example, Gale's [14] justification for considering an infinite-horizon program, one that surely was in the mind of Ramsey, and goes back to [45].

[10] In the context of Footnote 3, the word "turnpike" is being used as a noun.

[11] For this qualitative–quantitative distinction, we follow Pietsch's monumental history of Banach spaces; see [44].

[12] We shall have occasion to return to this blanket claim in the conclusion to this essay, and qualify it; see Footnote 36 below and the text it footnotes.

[13] See the bibliography of [6] and [3] for references to, and the reliance on, this classical literature even in the context of the particular subject-matter we discuss here.

1958 Dosso text [11]. In a brief and cursory footnote [4, Footnote 5; p. 543], he defers to McKenzie (1976) for a history of the term and to the survey of the attendant literature.

And so, as a second to attempt to get at an overview of the theory, one can perhaps proceed to Fischer's *Palgrave* 1987 entry[14] on Paul Samuelson in [13]. In his fourth section on the categorization of Samuelson's work in capital theory, Fischer writes:

> The theory of capital and growth sections of the first four volumes of *CSP* account for 38 papers, the largest single category.[15] Although capital theory is the branch of economics most vulnerable to Samuelson's comparative technical advantage, and although both his earliest 1937 papers are placed in that category in *CSP* (I:17, I:20), the output in this area is concentrated in *CSP* (III) covering the years from 1965 to 1970. Solow (1983) [53] provides a fine review of this part of Samuelson's research, some of which he co-authored. As Solow (1983) notes, much of the capital theory in *CSP* is related to developments in Dorfman, Samuelson and Solow (1958), which itself grew out of a 1949 Samuelson three-part memorandum for the Rand Corporation.

And as is well-known, especially to readers of Lionel McKenzie, the formal study of the turnpike theorem – also referred to as the "turnpike conjecture" at the time – was launched with the publication of Dosso.[16] But to get back to Fischer:

> Notable among the contributions is a variety of turnpike theorems. A turnpike theorem is conjectured in the 1949 Samuelson memorandum, and fully worked out in the 1958 volume. The theorem states that for any accumulation programme, starting from an initial vector of capital goods, and with specified terminal conditions, as the horizon lengthens the optimal programme spends an increasing proportion of its time near the von Neumann ray; more generally in

[14] This is reprinted without change in [12].

[15] Fischer uses the abbreviation *CSP* to refer to the five volumes of Samuelson's *Collected Papers* with the Roman numeral referring to the particular volume, and the Arabic numeral referring to the chapter in it.

[16] See McKenzie [34] for a detailed discussion to Dosso; also Mckenzie's 1963 papers [31, 32] and his 1987 *Palgrave* entry. The 1963 papers are masterly introductions to the pre-Ramsey literature, and devoted solely to what is being termed the "classical turnpike theorem" in this essay. Their introductions are useful complements to Fischer's "fully worked out" phrase in the quotation below. Also see McKenzie's description of the "Hicks' pilgrimage" in [31, Footnote 1], and a terminological clarification of the term *balanced growth* in the subsequent footnote.

> problems with intermediate consumption, the economy spends time near the modified golden rule. Several of the papers in the capital and growth section of *CSP*(III) contain turnpike theorems. A periodic turnpike result is reported in 1976, *CSP*(IV:224).

By seeing all the results from 1949 to the 1970s under a unified perspective, Fischer's entry then glosses over two points: first, an analytical turn, and a shift of emphasis, in the years 1965 to 1967 from a performance functional depending solely on the terminal capital stock of a finite-horizon allocation problem, an emphasis deriving from von Neumann's paper [55],[17] to one depending on the aggregate felicity obtained over the *entire* planning period, be it finite or infinite, deriving from Ramsey's paper [45]; and second, the dissonance between the undiscounted and discounted settings, a dissonance hinging precisely on the procedure of aggregation that is involved in the first shift. What Fischer does keep clear, however, is the linguistic subscription to what is being termed classical turnpike theory in this essay: he does not discuss asymptotic stability or include it within his outline of the subject.

Our attempt to get an up-to-date overview of turnpike theory began with Bewley's 2007 text, and moved on to Fischer's 1987 *Palgrave* entry. We now make a third attempt by going to the 2008 *Palgrave* entry of Brock–Dechert, experts who have also put their own particular stamp on the subject.

> A key property of the one-sector model that promotes its use in real business cycle applications as well as intertemporal general equilibrium asset pricing applications is the stochastic analog of the turnpike theorem. This theorem states that optimal capital stock and optimal consumption converge in a stochastic sense to a unique stochastic limit under standard assumptions of concavity of the payoff function (for example, the planner's preferences) and of the production function and modest assumptions on the structure of the stochastic shocks. It is much more difficult to obtain such results for general multisector stochastic models, and even for deterministic versions of those models.

Thus, classical turnpike theory is being ruled out from the outset in favor of asymptotic stability.[18] And in this context, Brock–Dechert consider the undiscounted case first, and write:

> The basic idea behind these results, called 'turnpike' results, is to first observe that, if the discount rate on the future is zero, the dynamic optimization problem will attempt to maximize a long-run

[17] See the references in the papers cited in Footnote 16 above.

[18] Thus we now have three different, albeit overlapping definitions of the term "turnpike theory."

> 'static' objective in order to avoid infinite 'value loss' if it failed to do so. Making this intuition mathematically precise requires introduction of a partial ordering called the 'overtaking ordering' and making assumptions on the objective function and the dynamics so that avoidance of infinite value loss results in convergence of the optimal quantities to a unique long-run limit.

In keeping with the thrust of this essay, the point of course is how to parlay the undiscounted results, and the associated mathematical techniques, into results for the discounted case. As we shall see in the sequel, this is not possible in the large, and one has to be satisfied with the local case. To quote Brock–Dechert again,

> Once one has results well in hand for the case of zero discounting on the future, intuition suggests that there should be a notion of 'continuity' that would enable one to prove that, if the discount rate is close enough to zero, convergence would still hold. Unfortunately, turning such intuition into precise mathematics turns out to be rather difficult (see McKenzie (1986, 2002), for deterministic literature and Arkin–Evstigneev (1987), the papers in Dechert (2001) [10], and Marimon (1989) [30] for the stochastic case).

And so, even if one confines oneself to the deterministic literature, as we do in this essay,[19] one would think that a natural point to begin would be McKenzie's 1987 and 2002 texts [38, 40], and then to move on to relevant papers in the current scholarly literature.[20] It is worthy of note in this context that there is no chapter on "turnpike theory in the 2006 Handbook [9]. However, a paper that is especially relevant is Mitra's 2005 "complete characterization of the turnpike property of optimal paths in the (reduced form) aggregative model of intertemporal allocation," [41], but in the very first sentence of his definitive paper, Mitra defines the "turnpike property" as equivalent to "global asymptotic stability of a non-trivial stationary optimal stock".[21] In sum, this literature is not always clear from the viewpoint of the

---

[19] This stochastic literature is huge, and deserving of its own overview. Our basic point is that the deterministic results ought to be antecedent to it as a matter of logical priority. At any rate, the 2008 *Palgrave* entry of Brock–Dechert is again worth quoting: "infinite horizon stochastic multisector models are also basic in constructing econometrically tractable models to use in analysing data. Here, especially, is where stochastic versions of the turnpike theorem (explained below) are used. For example, it is used to justify use of laws of large numbers and central limit theorems in econometric time-series applications."

[20] See [6, Chap. 6] for a comparative results in continuous time.

[21] Given the clarity of the phrase, one is led to inquire into the need for the turnpike terminology in the first place. It is almost as if it has been endowed with incantatory

distinction between classical turnpike theory and the local or global asymptotic stability of optimal programs. In any case, as the 2008 *Palgrave* entries attest, none of the theorems are simple enough to be presented without a detailed discussion of the hypotheses and the conclusions they engender, a task that we pursue here.

The outline of this essay is then as follows. In Sect. 2, we consider relevant results from the 1991 text by Carlson et al. [6] that considers both classical turnpike theory as well as the qualitative behavior of solutions to optimal control problems over an unbounded time interval. Given the primary integrative motivation behind this essay, the fact that this text confines itself to continuous time, and has both the calculus of variations and Pontryagin's maximum principle as background, is an added advantage. We supplement the discussion by a recent survey paper of Rockafellar's [46]. In Sect. 3 we present the model in discrete time. We follow the economic literature but also [3, 59]. In Sect. 4, we turn to the results on asymptotic stability based on the "smoothness hypotheses" on felicity functions as presented in the initial 1976 paper of Scheinkman's [52], and the subsequent extensions by Araujo–Scheinkman and McKenzie in the discrete-time framework; see [1, Definition 4.2]. These results constitute a remarkably successful application of the implicit function theorem for ℓ_∞ through an exploitation of "smoothness hypotheses" on the felicity function.[22] Section 5 moves away from the smoothness hypothesis, and considers McKenzie's refinements of the interiority assumptions. We also present his version of the "neighborhood turnpike theorem." Sections 6 and 7 then turn to recent results on the RSS and MW models, and discuss them in the light of previous work. Section 8 concludes the essay.

2. Relevant results in continuous time

The 1991 text by Carlson et al. [6] is scrupulous in maintaining the distinction between classical turnpike theory and the global asymptotic stability of solutions to infinite-horizon allocation problems,[23] and even though this essay is limited to allocation problems in discrete time, it is worthwhile to

power, for approval or for dismissal of the literature! We make a serious attempt in this essay to avoid the turnpike terminology when we can satisfactorily do so.

[22] See the comprehensive treatment of this theorem in Krantz–Parks [28].

[23] After a consideration of examples in Chap. 3, titled "asymptotic stability and the turnpike property in some simple control problems," the authors turn to the development of the general theory (in continuous-time) in in the subsequent four chapters. Their Sect. 6.7, and Chap. 7 is particularly relevant to the delineation of the discounted case that is being attempted in this essay.

consider their results. At the end of this section we shall supplement our discussion with a consideration of the recent survey of Rockafellar (2009), and his reliance on his own work, as well as that of Samuelson [50] and Levhari–Leviatan [29].

Consider the following problem subsequently referred to as Problem $\mathcal{P}_c$.

$$\max J_T(x_o, w(\cdot)) = \int_0^T f_0(x(t), w(t))dt \text{ subject to } \dot{x}(t) = f(x(t), w(t)) \text{ and } x(0) = x_o,$$

where $f(\cdot, \cdot) : \mathbb{R}^n \times R^m \longrightarrow \mathbb{R}^n$ is a continuous mapping with respect to both arguments $x(t)$ and $w(t)$, $W(x) \subset \mathbb{R}^m$ is compact, and that the mapping $x \to W(x)$ is upper semicontinuous.[24] This problem was originally considered in 1956 by Samuelson–Solow [51] as a multi-sectoral generalization of the problem of Ramsey [45].

Definition 2.1. *A pair of functions* $(x(\cdot), w(\cdot)) : [0, \infty) \longrightarrow \mathbb{R}^m$ *is said to be admissible if* $x(\cdot)$ *is absolutely continuous,* $w(\cdot)$ *is measurable,* $\dot{x}(t) = f(x(\cdot), w(\cdot))$, $w(t) \in W(x(t))$ *almost everywhere in* $[0, \infty)$ *and* $x(0) = x_o$. *Given an admissible pair of functions* $(x(\cdot), w(\cdot))$, *we shall refer to* $x(\cdot)$ *as a trajectory from* x_o *and generated by the admissible control* $w(\cdot)$.

We now turn to optimality criteria for infinite-horizon problems.

Definition 2.2. *A trajectory* $x^*(\cdot)$ *from* x_o *is said to be maximal if for any other trajectory from* x_o,

$$\liminf_{T\to\infty} \int_{t=0}^T J_T(x_o, w(\cdot)) - J_T(x_o, w^*(\cdot)) \leq 0.$$

A program is optimal if the lim sup *operator above is substituted for* lim inf.

The reader should note that *maximal* trajectories are referred to in the mathematical literature as *weakly overtaking optimal*, and *optimal* trajectories as *overtaking optimal* trajectories.[25]

For the Problem $\mathcal{P}_c$, let $(\hat{x}, \hat{w})$ be the solution to the following associated static problem:

$$\max f_o(x, w) \text{ such that } f(x, w) = 0 \text{ and } w \in W(x). \tag{1}$$

[24] This is the basic autonomous system considered in [6, Sect. 4.2] which should be referred to for the definitions and for simple examples. Note, however, that by following this text, we are not imposing non-negativity constraints of the state variables x.

[25] See [6, Definition 1.2] and the discussion following it. Rockafellar [46] also considers *optimal* trajectories where the inequality is strict and refers to them as satisfying a "strong" overtaking property.

Assumption 2.1. *Let (i)* $\hat{x}$ *be a unique solution to the stationary problem* (1), *(ii) there exist* $\hat{p} \in \mathbb{R}^n$ *such that* $f_o(\hat{x}, \hat{w}) = \max\{f_o(x, w) + \hat{p}f(x, w) : x \in \mathbb{R}^n, w \in W(x)\}$, *and (iii) for any* $\varepsilon > 0$, *there exists* $\delta > 0$ *such that*

$$||x - \hat{X}(\hat{x}, \hat{p})|| > \varepsilon \Longrightarrow \text{for all } w \in W(x),\ f_o(\hat{x}, \hat{w}) > (f_o(x, w) + \hat{p}f(x, w)) + \delta,$$

where $\hat{X}(\hat{x}, \hat{p}) \equiv \{x :$ *there exists* $w \in W(x)$ *with* $f_o(x, w) + \hat{p}f(x, w) = f_o(\hat{x}, \hat{w})\}$.

We can now present a basic result on the asymptotic stability of the solutions to the Problem $\mathcal{P}_c$.

Theorem 2.1. *Under Assumption 2.1 for the Problem* $\mathcal{P}_c$, *for any initial state* x_o *such that* $\hat{x}$ *is reachable in a finite time* $T(x_o)$, *any bounded maximal trajectory starting from* x_o *must satisfy* $\lim_{T \to \infty} x^*(t) \in \hat{X}(\hat{x}, \hat{p})$.

This theorem is presented in [6, Sect. 4.3] under the title "convergence toward the von Neumann set for weakly overtaking trajectories", and it is emphasized that the "classical maximum principle does not necessarily hold for the class" of functions for which the theorem is stated. In particular no differentiability assumptions on the function f_0 and f are made. And in the next section, the authors consider the "optimal control problem with fixed terminal time and prove a general turnpike theorem under the same assumptions which guarantee asymptotic stability of weak overtaking trajectories."

Theorem 2.2. *Under Assumption 2.1 for the Problem* $\mathcal{P}_c$, *for any initial state* x_o *such that* $\hat{x}$ *is reachable in a finite time* $T_1(x_o)$, *and any terminal state* x_T *reachable from* $\hat{x}$ *in a finite time* $T_2(x_T)$, *for any optimal trajectory* $x^*(\cdot)$ *and any* $T > T_1(x_o) + T_2(x_T)$, *the following property holds:*

$$\forall \varepsilon > 0 \text{ there exists } \tau(\varepsilon) > 0 \text{ such that } \mu[\{t \in [0, T] : ||x^*(t) - \hat{X}|| > \varepsilon\}] < \tau(\varepsilon),$$

where $\tau(\varepsilon)$ *does not depend on* T, *and* μ *denotes Lebesgue measure.*

In the remainder of Chap. 4 of [6], the authors consider the infinite-horizon version of the control problem $\mathcal{P}_c$, and using the results of Halkin [16] and Yano [56], present theorems on the existence and asymptotic stability of optimal and maximal trajectories. In particular, the results they present rely on what they refer to as the S property of the analogue of the von Neumann set. In Chap. 6 of their book [6], the authors consider both classical turnpike theory and local and global asymptotic stability of the solutions of the discounted version of Problem $\mathcal{P}_c$. We refer the reader to these results and satisfy ourselves with the following 2009 judgement of Rockafellar in [46].

The infinite-horizon problem we wish to investigate in general, as an extension of the finite-horizon problem

$$\max \int_0^T L_0(x(t), \dot{x}(t))dt \text{ subject to } x(0) = x_o, x(T) = x_T,$$

has the form

$$\max \int_0^\infty L_0(x(t), \dot{x}(t))e^{-\rho t}dt \text{ subject to } x(0) = x_o.$$

Do the results for the [first] problem have some counterpart for the [second]? A particular trouble-spot is what to make of the absence of a terminal constraint. Should one just allow $x(T)$ to do anything as $T \longrightarrow \infty$, or should a sort of infinite-horizon terminal constraint be imposed? Mathematical economists have had to contend with these issues unaided by much technical literature, and various difficulties have not been resolved to satisfaction. We need to understand better the infinite-horizon integral we are contemplating in the [second problem].

3. The basic model in discrete time

We now turn to the discrete-time setting, and consider the corresponding version of the Problem $\mathcal{P}_c$. We shall refer to it as the Problem $\mathcal{P}_d$, or $\mathcal{P}_d(x_0, \rho)$ for emphasis.

$$V_T(x_o, \rho) = \max \sum_0^\infty \rho^t u(x_t, x_{t+1}) \text{ subject to a given } x_0 = x_o$$

and where $u : \Omega \subset \mathbb{R}^{2n}_+ \longrightarrow \mathbb{R}$ and $0 < \rho < 1$. We shall also consider the case where $\rho = 1$ and refer to it as the undiscounted version of the Problem $\mathcal{P}_d$.

Araujo–Scheinkman [1] ascribe the original formulation of this problem to Samuelson–Solow [51], in the continuous-time version, and to Gale [14] and McKenzie [33] in the discrete-time version. They write:

> The reader more familiar with the "one-sector" growth theory would think of u as a "derived" utility function since there the utility function is defined in terms of consumption. There is no a priori reason to think of this as being a "derived" utility function since in many of the applications (e.g. the human capital theory) the utility function

does depend on the "state variables." Of course, the most general case is the one where the "original" utility depends on *both* "flow" and "stock" variables.[26]

The following basic concepts apply to the Problem $\mathcal{P}_d$ in either the undiscounted or discounted version.

Definition 3.1. *A program from x_o is a sequence $\{x(t)\}$ such that $x(0) = x_o$, and for all $t \in \mathbb{N}$, $(x(t), x(t+1)) \in \Omega$. A program $\{x(t)\}$ is simply a program from $x(0)$. A program $\{x(t)\}$ is called stationary if for all $t \in \mathbb{N}$, $(x(t)) = (x(t+1))$.*

We now turn to optimality criteria for a program. In the discounted case, with assumptions on Ω that ensure boundedness of programs, there is no issue of convergence of the performance functional.

Definition 3.2. *For all $0 < \rho < 1$, a program $\{x^*(t)\}$ from x_o is said to be optimal if*

$$\sum_{t=0}^{\infty} \rho^t [u(x(t), x(t+1)) - u(x^*(t), x^*(t+1))] \leq 0$$

for every program $\{x(t)\}$ from x_o. A stationary optimal program is a program that is stationary and optimal.

As first made explicit by Ramsey [45], the issue of the convergence of the performance functional becomes pressing only for the undiscounted case, which is to say, for the case $\rho = 1$. We can present the following analogue of Definition 2.2 for the discrete-time case.

Definition 3.3. *A program $\{x^*(t)\}$ from x_o is said to be maximal if for any other program from $\{x(t)\}$ from x_o,*

$$\liminf_{T\to\infty} \textstyle\sum_{t=0}^{T} u(x(t), x(t+1)) - u(x^*(t), x^*(t+1)) \leq 0.$$

A program is optimal if the lim sup *operator above is substituted for* lim inf.

We conclude this section by considering the discrete-time analogue of static problem considered above as (1).

$$\max\ u(x, x') \text{ such that } x' \geq x \text{ and } (x, x') \in \Omega. \tag{2}$$

[26] See [1, Footnote 2]. This is a rather prescient observation in the light of current interest in environmental economics. Perhaps such a more general model could be refereed to as the *MRSS* model: the *multi-sectoral Ramsey–Samuelson–Solow* model.

The interesting questions concern the discounted case where one solves the following problem:

$$\text{Find } (\hat{x}^{\rho}, \hat{x}^{\rho}) \in \Omega \text{ which solves } \max\ u(x, x') \text{ such that } x \leq (1-\rho)\hat{x}_{\rho} + \rho x' \text{ and } (x, x') \in \Omega. \quad (3)$$

Note that the above problem is a fixed point problem which reduces to the maximization problem represented in (2) only in the case $\rho = 1$. In the economic literature, the solution to (2) is referred to as the *golden-rule* stock, and to that of (3) as the *modified golden-rule* stock. For details as to these concepts, the reader can see McKenzie [38] and his references.[27] Also see the *weak* and *strong* turnpike distinction in [3].

4. Smooth felicities and asymptotic stability

In 1976, Scheinkman [52] assumes "smoothness hypotheses" on the felicity function u, and investigates whether the global asymptotic stability property of the optimal solutions to an undiscounted infinite-horizon problem can be translated to the optimal solutions of a discounted problem for "small" discount factors. This is to ask for a "neighborhood global stability theorem." His results were extended by Araujo–Scheinkman [1, 2] and McKenzie [35, 37, 38] to general perturbation results for arbitrary discount factors. These results constitute a remarkably successful application of the implicit function theorem for ℓ_∞, one that overcomes an important difficulty. The crucial issue can best be put in the words of Araujo–Scheinkman [1, p. 602].

> Since the implicit function theorem for such a space [the space of bounded sequences] reads exactly as in Euclidean spaces, the task becomes once again to study the inverse of a linear map. Given a linear map from a finite dimensional space into itself, the map is invertible if and only if it is one to one. This is no longer true in infinite dimensional spaces. Hence, concavity type restrictions . . . are not enough.

4.1. Scheinkman's "neighborhood" stability theorem

In an important paper in 1972, Samuelson [50], and following him, Levhari–Leviatan [29], had observed the saddle point property of optimal control motions. Rockafellar [46] discusses how the saddle-point property of the

[27] In the continuous-time case, the problem represented in (3) is referred to as an *implicit programming problem* in [6, Sect. 6.7].

associated Hamiltonian of the optimal control problem at the rest point translates into an "associated saddle point behavior of the Hamiltonian system in the dynamical sense" in the one-dimensional case, and asks:

> Could this somehow carry over to n-dimensions and in perturbation to positive discounting, at least if ρ isn't too high?

It is precisely an answer to this theorem that is provided in [52] as an application of the Hirsch–Pugh theorem.

We begin with the two basic assumptions of the texts [38, 40] that have now become standard for the general theory.

Assumption 4.1. *There exist $M > 0$ and $N < 1$ such that for all $(x, x') \in \Omega$, $||x|| < M$ implies $||x'|| < N||x||$.*

Assumption 4.2. *If $(x, x') \in \Omega$, then $(y, y') \in \Omega$ for all $y \geq x$, $0 \leq y' \leq x'$, and $u(y, y') \geq u(x, x')$ holds.*

Assumption 4.3. *There exists $(\bar{x}, \bar{x}') \in \Omega$ such that $\bar{x}' >> \bar{x}$. In this case, $\bar{x}$ is said to be expansible.*

We now present an assumption due to Brock [5] and fundamental to the theory; we already saw it in Assumption 2.1(i) above.

Assumption 4.4. *The golden-rule stock $\hat{x}$ is unique and expansible*

Assumption 4.5. *(i) $u(\cdot, \cdot)$ is C^3 in Int Ω and concave. (ii) The matrix*

$$\begin{bmatrix} u_{xx}(\hat{x}_\rho, \hat{x}_\rho) & u_{xx'}(\hat{x}_\rho, \hat{x}_\rho) \\ u_{x'x}(\hat{x}_\rho, \hat{x}_\rho) & u_{x'x'}(\hat{x}_\rho, \hat{x}_\rho) \end{bmatrix} \equiv \begin{bmatrix} A_\rho & B_\rho \\ B'_\rho & C_\rho \end{bmatrix}$$

is negative definite at $(\hat{x}, \hat{x})$. (iii) All programs are interior programs so that the Euler difference equation system

$$u_{x'}(x(t-1), x(t)) + \rho u_x(x(t), x(t+1)) = 0 \text{ for all } t \geq 1 \tag{4}$$

is satisfied for an optimal program. (iv) The characteristic equation of the Euler difference equation system (4) does not have a zero root. (v) There exists $\rho_\ell < 1$ such that for all $\rho \geq \rho_\ell$, there exists an expansible stationary program $\underline{x}$ such that $\hat{x}^\rho >> \underline{x}$.

Theorem 4.1 (Scheinkman). *Under Assumptions 4.1, 4.2, 4.3, 4.4 and 4.5, for any expansible $\underline{x}$, there exists $\bar{\rho} < 1$ such that $1 \geq \rho \geq \bar{\rho}$ and $x_o \geq \underline{x}$ implies that there exists an optimal program $x(t, x_o, \rho)$ such that $\lim_{t\to\infty} x(t, x_o, \rho) = \hat{x}^\rho$ where $\hat{x}^\rho$ is the unique modified golden-rule stock associated with ρ.*

The reader is referred to [37, pp. 347–350] for a description of Scheinkman's theorem, and to [52] for its proof. The "visit lemma" in the proof does not require any smoothness hypotheses, and plays an important role in [22, 25], as discussed below.

4.2. Refinements of the "neighborhood" theorem

We now turn to the results of Araujo–Scheinkman [1]. We begin by providing an alternative to Assumption 4.5 which will be used as the Standing Hypothesis for all of the remaining results in this section.

Assumption 4.6. *The felicity function u is strictly concave in Ω and C^2, with D^2u negative definite in Int (Ω).*

Definition 4.1. *A program $\{x(t)\}$ is said to be a regular program if there exists $\varepsilon > 0$ such that*

$$\inf\{d_H[(x_t, x_{t+1}), \partial\Omega] : t = 0, 1, \cdots\} > 0,$$

where d_H denotes the Hausdorff distance, and $\partial\Omega$ the boundary of Ω.

We shall denote by the first projection of Ω by $\Pi_1(\Omega)$ and by $\mathcal{S}(\mathcal{S}^r)$ the set of all pairs of stocks and discount factors (x, ρ), $x \in \Pi_1(\Omega)$, $\rho \in (0, 1)$ for which there exists a solution (regular solution) to $\mathcal{P}_d(x, \rho)$. The optimal program function $x : \mathcal{S} \longrightarrow \ell^n_\infty$, $x(x_o, \rho)$ is a solution to $\mathcal{P}_d(x_o, \rho)$. Let

$$\tau = \{(x, \{x(t)\}) : \{x(t)\} \text{ is a regular program from } x\}.$$

We shall need the mapping

$$\Phi : \tau \times (0, 1) \longrightarrow \ell^n_\infty \text{ by } [\Phi(x_o, \{x(t)\}, \rho)]_t = u_{x'}(x(t-1), x(t)) \\ +\rho u_x(x(t), x(t+1)).$$

Note that for a given ρ, Φ maps a regular program into the left hand side of the Euler difference equation system (4). A crucial assumption of the analysis is the following[28]

Assumption 4.7. *$T = D_x\Phi(x_o, \{x(t)\}, \rho)$ has dominant diagonal blocks.*

Theorem 4.2. *Under Assumption 4.6, for any $(x_o, \rho) \in \mathcal{S}^r$ satisfying Assumption 4.7, there exists $\varepsilon > 0$ such that $B_\varepsilon(x_o, \rho) \subset \mathcal{S}^r$ and $x : B_\varepsilon(x_o, \rho) \longrightarrow \ell^n_\infty$ is C^1.*

We shall now need the following concepts.

Definition 4.2. *For any $\rho \in (0, 1)$, a modified golden-rule stock $\hat{x}^\rho$ is said to be locally asymptotically stable if there exists $\varepsilon > 0$ such that*

$$x_o \in \Pi_1(\Omega),\ |x_o - \hat{x}^\rho| < \varepsilon \Longrightarrow \lim_{t\to\infty} x(t, x_o, \rho) = \hat{x}^\rho.$$

[28] The reader is referred to [1, Definition 2.4] for the precise definition. We also note that the symbol x is being used in two senses: the value of a stock at a particular time, as well as a function from the initial stock and the discount factor to ℓ^n_∞.

Definition 4.3. *For any $X \subset \Pi_1(Int\ \Omega)$ and a $\rho \in (0, 1)$, a modified golden-rule stock $\hat{x}^\rho$ is said to be X-globally asymptotically stable at ρ if*

$$x_o \in X,\ (x_o, \rho) \in \mathcal{S}^r \implies \lim_{t\to\infty} x(t, x_o, \rho) = \hat{x}^\rho.$$

If X equals $\Pi_1(Int\ \Omega)$ in the above definition, $\hat{x}^\rho$ is said to be globally asymptotically stable at ρ.

The next result attests to the power of the continuity hypothesis in the strengthening the local asymptotic stability of a modified golden-rule program to a larger set of initial stocks.

Theorem 4.3. *For any $\rho \in (0, 1)$ and any connected $X \subset \Pi_1(\Omega)$ such that (i) $(x_o, \rho) \in \mathcal{S}^r$ for all $x_o \in X$, (ii) $x(\cdot, \rho) : X \longrightarrow \ell^n_\infty$ is continuous, and (iii) $\hat{x}^\rho$ is locally asymptotically stable, $\hat{x}^\rho$ is X-globally asymptotically stable.*

The question then devolves to a result on global asymptotic stability without the continuity assumption, which is to ask for a result with assumptions on the primitives. We present this next.

Theorem 4.4. *Under Assumption 4.6, for any $\rho \in (0, 1)$ such that (i) a modified golden-rule stock $(x^\rho, x^\rho) \in Int\ (\Omega)$, (ii) $(x_o, \rho) \in \mathcal{S}^r$, (iii) Assumption 4.7 holds for all $x_o \in Int\ (\Omega)$, x^ρ is globally asymptotically stable.*

In order to develop the next assumption, we shall abbreviate our notation to denote by u_1 the vector of partial derivatives of u with respect to the first n coordinates of $\mathbb{R}^{2n}$, and by u_2 the vector of partial derivatives of u with respect to the last n coordinates of $\mathbb{R}^{2n}$. Correspondingly, we shall denote by $u_{i,j}$, $i = 1, 2$; $j = 1, 2$, the matrices of partial derivatives of u_i, $i = 1, 2$. Thus, in terms of the earlier notation, u_{21} denotes $u_{x'x}$. Finally, we shall denote by u^ρ_{ij} the matrix $u_{x'x}(\hat{x}^\rho, \hat{x}^r ho)$.

Assumption 4.8. *The modified golden-rule stock $(\hat{x}^\rho, \hat{x}^\rho) \in Int\ (\Omega)$, $det\ u_{x'x}(\hat{x}^\rho, \hat{x}^\rho) \neq 0$, and there exist n roots $\lambda_1, \cdots, \lambda_n$, all of modulus less than one, of the characteristic equation*

$$det\ (u^\rho_{21} + \lambda[u^\rho_{22} + \rho u^\rho_{11}] + \lambda^2 \rho u^\rho_{12}) = 0.$$

Theorem 4.5 (Araujo–Scheinkman). *Under Assumption 4.6, for any $\rho \in (0, 1)$ satisfying Assumption 4.8, and $x_o \in Int\ (\Omega)$ such that $(x_o, \rho) \in \mathcal{S}^r$ and $\lim_{t\to\infty} x(t, x_o, p) = \hat{x}^\rho$, $T(x_o, \rho)$ is an isomorphism. Consequently, there exists $\varepsilon > 0$ such that $B_\varepsilon(x_o, \rho) \subset \mathcal{S}^r$ and the restriction of the optimal program function x to $B_\varepsilon(x_o, \rho)$ is C^1.*

By strengthening the asymptotic stability of the modified golden-rule stock x^{ρ} to global asymptotic stability in Theorem 4.5, Araujo–Scheinkman state two important corollaries of the result for the given ρ: (i) that the C^1 property of the optimal program holds for all $x_o \in \text{Int}(\Omega)$, (ii) that on the optimal program, the optimal policy function from the stock in one period to that in the next is also C^1. But the *piéce de resistance* is the following theorem in the neighborhood of the discount factor.

Theorem 4.6. *Under Assumption 4.6, for any* $\bar{\rho} \in (0, 1)$ *satisfying Assumption 4.8,* $x_o \in Int(\Omega)$, $(x_o, \bar{\rho}) \in S^r$, *the modified golden-rule stock* $\hat{x}^{\bar{\rho}}$ *is globally asymptotically stable, and any compact set* $X \subset \Pi_1(Int\ \Omega)$, *there exists* $\varepsilon > 0$ *such that for any* $\rho \in B_{\varepsilon}(\bar{\rho})$, x^{ρ} *is* X*-asymptotically stable at* ρ.

Araujo–Scheinkman remark that Theorem 4.6 can be strengthened to a result on global asymptotic stability if the Visiting Lemma holds. This is to say that for any $x_o \in \text{Int}(\Omega)$ and any $\rho \in B_{\varepsilon}(\bar{\rho})$, and for a suitable set of stocks K, there exists $t(x_o, \rho)$ such that $x_{t(x_o,\rho)} \in K$.

5. The theory in non-smooth environments

In the chapter on *Competitive Equilibrium over Time,* McKenzie [40] presents results for a "generalized Ramsey model", and his work is distinguished by its reliance on a bounded assumption on the technology Ω, and a joint assumption on the pair (u, Ω) guaranteeing *free-disposal* and *monotonicity.*

Under Assumptions 4.1 and 4.2, he presents theorems for both the discounted and undiscounted cases, seeing the latter as logically antecedent to the former.

5.1. Theorems under refined interiority assumptions

We can now present a classical turnpike theorem, Card(A) denoting the cardinality of a finite set A.

Theorem 5.1. *Let Assumptions 4.1, 4.2 and 4.3 hold and that* $\hat{x}$ *is expansible. Then, there exists* $\varepsilon > 0$ *such that for all* $T > L$, *and all optimal programs* $\{x_t\}_{t=0}^{t=T}$ *such that* x_0 *and* x_T *are expansible,*

$$Card\big\{i \in [0, \cdots, T-1] : \|x(t) - X(\hat{x}, \hat{p})\| > \varepsilon\big\} \leq L.$$

This is presented in [40, Theorem 3, Chap. 7] and proved there.

Mckenzie introduces his other theorem for the undiscounted Ramsey setting as "an asymptotic theorem for infinite optimal paths." For this, he needs a particular set of initial stocks K from which there start programs $\{x(t)\}$ such that

$$\liminf_{T\to\infty} \sum_{t=0}^{T} u(x(t), x(t+1)) - u(x^*(t), x^*(t+1)) \geq -\infty.$$

These are the *good* programs of Gale [14] and Brock [5]; see Assumptions 2.1 and 4.4.

Theorem 5.2. *Let Assumptions 4.1, 4.2 and 4.3 hold and that $\hat{x}$ is unique. Let $\{x(t)\}$ be an optimal program that starts from a stock in the interior of K. Then, for any $\varepsilon > 0$, there exists T_o such that $||x(t) - \hat{x}|| < \varepsilon$ for all $t \geq T_o$.*

5.2. McKenzie's "neighborhood turnpike" theorem

We shall now make assumptions on the discount factor. We shall need the following additional notation. Let $D = \{x \in \mathbb{R}^n_+ : (x, x) \in \Omega\}$, and

$$D_\rho = \{x \in D : u(x, x) \geq u(\bar{x}, \bar{x}') \text{ for all } (\bar{x}, \bar{x}') \in \Omega \text{ such that } \rho\bar{x}' >> x\}.$$

Before introducing the next assumption, we shall need McKenzie's definition of *uniform strict concavity* of a function taken from [40, Sect. 7.6].

Definition 5.1. *The utility function u is said to be uniformly strictly concave over D_ρ if for any $bx \in D_\rho$, and any $(w, z) \in \Omega$, the following holds: For any $\varepsilon > 0$, there exists $\delta > 0$ such that*

$$|(x, x) - (w, z)| > \varepsilon \Longrightarrow u\left(\frac{1}{2}(x, x) + \frac{1}{2}(w, z)\right) - \frac{1}{2}(u(x, x) + u(w, z)) > \delta.$$

We can now present

Assumption 5.1. *There exists $\rho_\ell < 1$ and $(\bar{x}, \bar{x}') \in \Omega$ such that $\rho_\ell \bar{x}' >> \bar{x}$. We assume $\rho_\ell \leq \rho \leq 1$, u is uniformly strictly concave over D_ρ, and that D_ρ is in the relative interior of D.*

Assumption 5.2. *$\hat{x}$ is expansible which is to say that there exists $(\hat{x}, x') \in \Omega$ such that $x' >> \hat{x}$. Furthermore D_ρ is in the relative interior of D.*

Definition 5.2. *We shall say that an initial stock $x_o \in \mathbb{R}^n_+$ is sufficient if there exists a finite program $\{x_t\}_{t=0}^{t=T}$ such that x_T is expansible.*

We can now present McKenzie's so-called "neighborhood turnpike theorem" for any expansible $\hat{x}$.

Theorem 5.3. *Under Assumptions 4.1 and 4.2, for any sufficient $x_o \in \mathbb{R}^n$, and any ρ satisfying Assumptions 5.1 and 5.2, and any $\varepsilon > 0$, there exists $1 > \bar{\rho} > 0$ such that $\hat{x}^\rho$ is asymptotically stable at (ρ, x_o) for all $\bar{\rho} \leq \rho \leq 1$.*

For the proof, the reader is referred to that of Theorem 6 in [40, Sect. 7.6]. Note that in [36, Theorem 4] McKenzie presents another theorem which he introduces as the "neighborhood turnpike" and refers to it as the "basic result of the subject." However, he does not report this theorem in his text [40]. It is clear that further comparative work needs to be done.

6. Particular undiscounted cases: the RSS model

There is by now an extensive literature on the so-called RSS model, and we refer the reader to the relevant references: for its underpinnings and interpretation as a model in development planning and the antecedent literature due to Robinson, Solow and Srinivasan, see [17]; for details of the important two-sector RSS case, see [18, 20] and their references; for the continuous time analysis, see [19, 54]. Our main concern here is to bring out the fact that the RSS model is a particular case of the model spelt out in Sect. 3.[29]

The model requires the total labor force to be stationary, perfectly divisible and positive, and thereby normalized to unity. There are n types of machines, produced using only labor, with $a_i > 0$ units of labor are needed to produce one unit of a perfectly divisible machine of type i, for $i = 1, \ldots, n$. Let a denote the n-vector $(a_1, \cdots, a_n)$. The *state* of the system will be represented by a point $x \in \mathbb{R}^n_+$, where x_i represents the current stock of machines of type $i = 1, \ldots, n$. If x represents the stock of machines today, x' the stock tomorrow and $d \in (0, 1)$ the common rate of depreciation, then $z = x' - (1 - d)x$ stands for the number of machines produced during the period. The constraints $z \geq 0$ and $az \leq 1$ represent respectively the irreversibility of investment and the maximum labor available, where az denotes inner product. With these definitions, the *transition possibility set* is given by

$$\Omega = \{(x, x') \in \mathbb{R}^n_+ \times \mathbb{R}^n_+ : x' - (1-d)x \geq 0 \text{ and } a(x' - (1-d)x) \leq 1\}. \quad (5)$$

Given the pair $(x, x') \in \Omega$ and denoting by e the vector $(1, 1, \cdots, 1) \in \mathbb{R}^n$, the stock of machines that may be devoted to the consumption goods sector is given by the correspondence

$$\Lambda(x, x') = \{y \in \mathbb{R}^n : 0 \leq y \leq x \text{ and } ey \leq 1 - a(x' - (1 - d)x)\}. \quad (6)$$

[29] We leave it to the reader as an exercise to show that the continuous-time version of the model, as presented in [19, 54] and their references, is a particular case of the model spelt out in Sect. 2.

One unit of labor together with one unit of machine of type $i = 1, \ldots, n$, can produce $b_i > 0$ units of a single consumption good, thus, defining the output-coefficients vector $b = (b_1, \ldots, b_n) \in \mathbb{R}^n_+$, the total good production is given by by, with $y \in \Lambda(x, x')$.

Welfare is derived only from the consumption good and is represented by a felicity function $w : \mathbb{R}_+ \to \mathbb{R}_+$ that is assumed to be continuous, strictly increasing, concave and differentiable. In terms of the triple (Ω, u, x_o) pertaining to the general theory, and mentioned in the first sentence of the introduction, we can now define

$$u(x, x') = \arg\max\{w(by) : y \in \Lambda(x, x')\} \text{ for all } (x, x') \in \Omega.$$

In the results reported in this section, we shall make the following assumption:

Assumption 6.1. *There exists $\sigma \in \{1, \ldots, n\}$ such that for all $i \in \{1, \ldots, n\} \setminus \{\sigma\}$, $c_\sigma > c_i$, where $c_i = b_i/(1 + da_i)$, $i = 1, \ldots, n$.*

This is a fundamental assumption that guarantees a unique golden-rule stock, and thereby a unique von-Neumann ray. This assumption goes back to Brock [5], and for a detailed discussion in the context of the RSS model, the reader is referred to [17]; also see [57].

We shall also work with the basic parameter (and attendant notation) that was identified in the discrete-time context [17], and was totally missed in the continuous-time setting originally studied by Stiglitz [54].

$$\xi_i = (1/a_i) - (1 - d) \text{ for all } i = 1, \cdots, n. \tag{7}$$

Even though a special case of the general theory, the model does not fulfill Assumptions 4.5 and 4.6 that the theory requires.

6.1. A theorem on asymptotic stability

We now present a basic result on the asymptotic stability

Theorem 6.1. *Assume that (i) w is strictly concave, or that (ii) $\xi_\sigma \neq 1$. Let $M_0, \epsilon > 0$. Then there exists a natural number T_0 such that for each optimal program $\{x(t), y(t)\}_{t=0}^{\infty}$ satisfying $x(0) \leq M_0 e$ and each integer $t \geq T_0$*

$$||x(t) - \widehat{x}||,\ ||y(t) - \widehat{x}|| \leq \epsilon.$$

Note that the time period T_o is does not depend on the initial stock, lying in the given range of initial stocks, from which the optimal program starts. For a proof, the reader is referred to [24].

6.2. Classical turnpike theorems

The basic results are as follows.

Theorem 6.2. *Let M, ϵ be positive numbers and $\Gamma \in (0,1)$. Then there exists a natural number L such that for each integer $T > L$, each $z_0, z_1 \in R^n_+$ satisfying $z_0 \le Me$ and $az_1 \le \Gamma d^{-1}$ and each program $(\{x(t)\}_{t=0}^{T}, \{y(t)\}_{t=0}^{T-1})$ which satisfies*

$$x(0) = z_0, \; x(T) \ge z_1, \quad \sum_{t=0}^{T-1} w(by(t)) \ge U(z_0, z_1, 0, T) - M,$$

the following inequality holds:

$$Card\{i \in \{0, \ldots, T-1\} : \max\{||x(t) - \widehat{x}||, \; ||y(t) - \widehat{x}||\} > \epsilon\} \le L.$$

Theorem 6.3. *Let M, ϵ be positive numbers and $\Gamma \in (0,1)$. Then there exist a natural number L and a positive number γ such that for each integer $T > 2L$, each $z_0, z_1 \in R^n_+$ satisfying $z_0 \le Me$ and $az_1 \le \Gamma d^{-1}$ and each program $(\{x(t)\}_{t=0}^{T}, \{y(t)\}_{t=0}^{T-1})$ which satisfies*

$$x(0) = z_0, \; x(T) \ge z_1, \quad \sum_{t=0}^{T-1} w(by(t)) \ge U(z_0, z_1, 0, T) - \gamma,$$

there are integers τ_1, τ_2 such that $\tau_1 \in [0, L], \; \tau_2 \in [T-L, T]$,

$||x(t) - \widehat{x}||, \; ||y(t) - \widehat{x}|| \le \epsilon$ for all $t = \tau_1, \ldots, \tau_2 - 1$ and $||x(\tau_2) - \widehat{x}|| \le \epsilon$.

Moreover if $||x(0) - \widehat{x}|| \le \gamma$ then $\tau_1 = 0$.

For proofs and notation, the reader is referred to [25]. This reference also derives the asymptotic stability result as a corollary of these theorems. Many turnpike results of this kind motivated by non-economic applications are collected in [38].

7. Particular undiscounted cases: the MW model

There is by now a vigorously-developing literature on the so-called MW model, and we refer the reader to the relevant references: the pioneering papers are of Mitra–Wan [42, 43] subsequent to those of Faustman and Samuelson.[30] As in Sect. 6, our main concern here is to bring out the fact that the MW model is a particular case of the model spelt out in Sect. 3.[31]

[30] For the recent papers, see [47, 48] and [21, 22] and their references.

[31] The continuous-time version of the model is difficult, and no longer an easily formulated particular version of the model spelt out in Sect. 2; see [42, 43] for the early formulations in the work of Kemp and Wan.

We shall work in the $(n-1)$-dimensional simplex $\Delta = \{x \in \mathbb{R}_+ : \sum_{i=1}^n x_i = 1\}$, let the supremum norm of x be $\|x\|_\infty$. Let the total surface be unity and n the age after which a tree dies or losses its economic value. We consider that the timber content per unit of area is related only to the age of the trees, through the biomass coefficient vector $b = (b_1, \cdots, b_n)$. We make an assumption analogous to Assumption 6.1 that:

Assumption 7.1. *There exists $\sigma = \{1, \cdots, n\}$ such that $b_\sigma/\sigma > b_i/i$ for all $i \in \{1, \cdots, n\} \setminus \{\sigma\}$.*

For each period $t \in \mathbb{N}$ we denote $x_i(t) \geq 0, i = 1, \cdots, n$ the surface occupied by trees of age i at time t. At every stage we must decide how much land to harvest of every age-class, $c(t) = (c_1(t), \cdots, c_n(t))$ where $c_i(t) \in [0, x_i(t)]$. As we know that after n a tree has no value, we assume that $c_n(t) = x_n(t)$ for all t. By the end of period $t+1$, the state will be exactly

$$x(t+1) = \left(\sum_{i=1}^n c_i(t), x_1(t) - c_1(t), \cdots, x_{n-1}(t) - c_{n-1}(t)\right).$$

Define the transition possibility set Ω as the collection of pairs $(x, x') \in \Delta \times \Delta$ such that it is possible to go from the state x in the current period (today) to the state of the forest x' in the next period (tomorrow) fulfilling relations. Formally,

$$\Omega = \{(x, x') \in \Delta \times \Delta : x_i \geq x'_{i+1} \text{ for all } i = 1, \cdots, n1\}$$

The vector of harvests needed to perform this transition is given by the function $\lambda : \Omega \to \mathbb{R}^n_+$, $\lambda(x, x') = (x_1 - x'_2, x_2 - x'_3, \cdots, x_{n-1} - x'_n, x_n)$. In addition, it is easy to see that $(x, x') \in \Omega \Leftrightarrow x, x' \in \Delta$ and $\lambda(x, x') \geq 0$.

The preferences of the planner are represented by a felicity function, $w : [0, \infty) \to \mathbb{R}$ which is assumed to be continuous, strictly increasing and concave. Define for any $(x, x') \in \Omega$ the function $u(x, x')$ as

$$u(x, x') = w(bc) \text{ where } c = \lambda(x, x')$$

This model is in fact an equivalent formulation of the one proposed in [42, 43],[32] and one that does not fulfill assumptions 4.2 and 4.4 of the general theory. The set Ω has no interior point in $\mathbb{R}^{2n}$, and the natural preorder does not apply to it. A reformulation fulfills Assumption 4.3 but not 4.4.

In the discounted case, the modified golden-rule stock is not asymptotically stable, see [47, 48]. However the validity of the turnpike theorems is still an open question. Here again Assumptions 4.6 and 4.8 are not fulfilled: non-interiority and non-concavity remains a thorny issue.

---

[32] For differences in the model as presented in [42, 43] and here, see [22].

7.1. A theorem on asymptotic stability

We first present an asymptotic stability result taken from [23], a reference that contains further discussion and proof.

Theorem 7.1. *Let $\varepsilon > 0$. There exists a natural number T_0 such that for each optimal program $\{x(t)\}$ the following inequality holds:*

$$\|x(t) - \hat{x}\|_\infty < \varepsilon \text{ for all } t \geq T_0.$$

For a proof of this theorem, the reader is referred to [23].

7.2. Classical turnpike results

Next, we present classical turnpike results taken from [22], again a reference that contains further discussion and proof.

Theorem 7.2. *Given $M > 0$ and $\varepsilon > 0$ there exists $L \in \mathbb{N}$ such that for all $T > L$ and each program $\{x(t)\}_{t=0}^T$ satisfying*

$$\sum_{t=0}^{T-1} w(bc(t)) \geq U\big(x(0), x(T), 0, T\big) - M,$$

$$Card\big\{i \in [0, \cdots, T-1] : \|x(t) - \hat{x}\| > \varepsilon\big\} \leq L.$$

Theorem 7.3. *Let $\varepsilon > 0$. Then there exist $L \in \mathbb{N}$ and $M > 0$ such that for all $T > 2L + n + \sigma$ and each program $\{x(t)\}_{t=0}^T$ satisfying*

$$\sum_{t=0}^{T-1} w(bc(t)) \geq U\big(x(0), x(T), 0, T\big) - M,$$

there are τ_1, τ_2 such that $\tau_1 \in [0, L]$, $\tau_2 \in [T - L, T]$ and $\|x(t) - \hat{x}\| \leq \varepsilon$ for all $t = \tau_1, \cdots, \tau_2$. Moreover, if $\|x(0) - \hat{x}\| \leq \varepsilon/n^2$ then $\tau_1 = 0$.

8. Conclusion

The alert reader has surely noticed that in the work presented above, we have been silent on the aggregative, one-sector setting, referred to as the RCK model in the introduction. We briefly remedy this deficiency here. As is well-known, the pioneering results on the discounted case in this setting are due to Cass [7, 8], Koopmans [26, 27] and Samuelson [49]. Indeed, Samuelson sketches the theorem in the last footnote of his 1965 paper as follows:

> Suppose society has a systematic subjective rate of time preference ρ, so that the integrand of $u(c)$ is replaced by $u(c)e^{-\rho t}$. Then a new "turnpike" is given by the root of $f'(x) = \rho$ rather than equal zero. Call this x^ρ. Then for T sufficiently large, for all (x_o, x_T), the optimal path $x(t)$ will remain indefinitely near to x^ρ an indefinitely large

> fraction of the time. The motions are not quite balanced catenaries, instead approximating to $k(t) = a_1 e^{\lambda_1 t} + a_2 e^{\lambda_2 t}, \lambda_1 > \lambda_2$. This constitutes an even more general Turnpike Theorem than the one elaborated here, and it reminds us that correct saving must depend on what is assumed about social time preference.[33]

At this point in the exposition we can connect to the epigraph of this essay taken from a neglected paper of Gale's [15].[34] Unlike Ramsey, Samuelson, Solow and Cass, Gale conducts the analysis in discrete-time, and shows that the results on the characterization of solutions of the RCK model are too good to be true: the classical turnpike theorem is valid without any restriction on the discount factor. In a recent paper [41], Mitra provides a synthesis of sufficient conditions for asymptotic stability of optimal programs in the RCK model, but has nothing to say on what we are referring to as classical turnpike theory in this essay.

In conclusion, and as emphasized in the introduction, the work presented above has been governed by two objectives: first, to encourage productive engagement across disciplinary boundaries; and second, to identify open problems for both disciplines. The point is to provide an overview that would benefit and bring together the two communities of economists and mathematicians: the former working in growth theory and economic dynamics, and the latter in dynamical systems and optimal control theory.[35] A terminological distinction between classical turnpike theory and the qualitative behavior of solutions to optimal control problems over an unbounded time interval, in either discrete or continuous time, certainly advances this objective.[36] Ironically, recent work of economists has concerned itself with the qualitative investigation of asymptotic stability of optimal control motions in discrete time, and it has been left to mathematicians to investigate classical turnpike theory in continuous time, with of course notable exceptions in both cases.[37] As regards the second objective, the aspiration remains for a generalized theory pertaining to the model due to Samuelson–Solow (1956), Gale (1967)

---

[33] If footnotes could be labeled "pioneering," surely this footnote (notationally modified) would be high on such a list. It is a fascinating exercise in the history of economic analysis to trace the work surveyed in Sect. 4 as an elaboration of this footnote.

[34] In the context of this epigraph, the reader should note that the Kuhn–Tucker theorem is now referred to as the Karush–Kuhn–Tucker theorem, and what Gale refers to as the *discount rate* is the *discount factor.* The first sentence of the epigraph is taken from page 308, the second from pp. 314–315 and the third from p. 310.

[35] We see Rockafellar's recent survey [46] as being kindred in this motivation.

[36] And so we end as we began: with the importance of having a clear and well-established terminology. See Footnote 12 above and the text it footnotes.

[37] See, for example, Carlson et al. [6], Arkin–Evstegneev [3] and Zaslavski [58], and their references, to the work of both economists and mathematicians.

and McKenzie (1968), one that moves seamlessly between the discounted and undiscounted cases, asymptotically implements the continuous-time results in terms of their discrete-time counterparts, and covers both the RSS and MW models, the two concrete instances in development planning and the economics of forestry considered above. As we have seen, non-interiority assumptions are endemic to these applications, and smoothness, strict concavity and corresponding curvature assumptions do not translate to corresponding assumptions on the reduced-form performance functionals in which they are phrased.

References

1. Araujo, A., Scheinkman, J.A.: Comparative dynamics and the turnpike property. Econometrica **45**, 601–620 (1977)
2. Araujo, A., Scheinkman, J.A.: Notes on comparative dynamics. In: Green, J.R., Scheinkman, J.A. (eds.) General Equilibrium, Growth and Trade, pp. 217–226. Academic, New York (1979)
3. Arkin, V., Evstigneev, I.: Stochastic Models of Control and Economic Dynamics. Academic, New York (1987)
4. Bewley, T.F.: General Equilibrium, Overlapping Generations Models and Optimal Growth Theory. Harvard University Press, Cambridge (2007)
5. Brock, W.A.: On existence of weakly maximal programmes in a multi-sector economy. Rev. Econ. Stud. **37**, 275–280 (1970)
6. Carlson, D.A., Haurie, A.B., Leizarowitz, A.: Infinite Horizon Optimal Control: Deterministic and Stochastic Systems. Springer, Berlin (1991)
7. Cass, D.: Optimum growth in an aggregative model of capital accumulation. Rev. Econ. Stud. **32**, 233–240 (1965)
8. Cass, D.: Optimum growth in an aggregative model of capital accumulation: a turnpike theorem. Econometrica **34**, 833–850 (1966)
9. Dana, R.A., Le Van, C., Mitra, T., Nishimura, K. (eds.): Handbook of Optimal Growth, vol. 1. Springer, Berlin (2006)
10. Dechert, W. (ed.): Growth Theory, Nonlinear Dynamics, and Economic Modelling: Scientific Essays of William Allen Brock. Edward Elgar, Cheltenham (2001)
11. Dorfman, R., Solow, R.M., Samuelson, P.A.: Linear Programming and Economic Analysis. McGraw-Hill, New York (1958)
12. Durlauf, S., Blume, L. (eds.): The New Palgrave, vols. 1–7. Macmillan, New York (2008)
13. Eatwell, J., Milgate, M., Newman, P.K. (eds.): The New Palgrave, vols. 1–4. Macmillan, New York (1987)

14. Gale, D.: On optimal development in a multi-sector economy. Rev. Econ. Stud. **34**, 1–18 (1967)
15. Gale, D.: Nonlinear duality and qualitative properties of optimal growth. In: Abadie, J. (ed.) Integer and Nonlinear Programming, pp. 309–319. North Holland, Amsterdam (1970)
16. Halkin, H.: Necessary conditions for optimal control problems with infinite horizon. Econometrica **42**, 267–273 (1974)
17. Khan, M.A., Mitra, T.: On choice of technique in the Robinson–Solow–Srinivasan model. Int. J. Econ. Theory **1**, 83–109 (2005)
18. Khan, M.A., Mitra, T.: Discounted optimal growth in the two-sector RSS model: a geometric investigation. Adv. Math. Econ. **8**, 349–381 (2006)
19. Khan, M.A., Mitra, T.: Optimal growth in the two-sector RSS model: a continuous time analysis. In: Proceedings of the Seventh Portugese Conference on Automatic Control, Lisboa, Portugal. Electronic publication (2006)
20. Khan, M.A., Mitra, T.: Optimal growth in a two-sector RSS model without discounting: a geometric investigation. Jpn. Econ. Rev. **58**, 191–225 (2007)
21. Khan, M.A., Piazza, A.: On the Mitra-Wan Forestry Model: A Unified Analysis. *Publicación Técnica CMM* No. 226. Paper presented at *Workshop in Economic Dynamics,* National University of Singapore, July 30–August 1, 2009 (2009)
22. Khan, M.A., Piazza, A.: Classical turnpike theory and the economics of forestry. J. Behav. Econ. Organ. (2010, forthcoming)
23. Khan, M.A., Piazza, A.: On uniform convergence of undiscounted optimal programs in the Mitra-Wan forestry model: the strictly concave case. Int. J. Econ. Theory **6**, 57–76 (2010)
24. Khan, M.A., Zaslavski, A.J.: On a uniform turnpike of the third kind in the Robinson–Solow–Srinivasan model. J. Econ. **92**, 137–166 (2006)
25. Khan, M.A., Zaslavski, A.J.: On two classical turnpike results for the Robinson–Solow–Srinivasan (RSS) model. Adv. Math. Econ. **13**, 47–97 (2010)
26. Koopmans, T.C.: On the concept of optimal economic growth. In: Pontificae Academiae Scientarum Scripia Veria (ed.) Economic Approach to Development Planning, pp. 225–287. North Holland, Amsterdam (1965)
27. Koopmans, T.C.: Intertemporal distribution and "optimal" aggregate economic growth. In: Fellner, W., et al. (eds.) Ten Economic Studies in the Tradition of Irving Fisher, pp. 95–126. Wiley, New York (1967)
28. Krantz, S.G., Parks, H.R.: The Implicit Function Theorem: History, Theory and Applications. Birkhäuser, Basel (1998)
29. Levhari, D., Leviatan, N.: Stability in the saddlepoint sense. J. Econ. Theory **4**, 88–93 (1972)

30. Marimon, R.: Stochastic turnpike property and stationary equilibrium. J. Econ. Theory **47**, 282–306 (1989)
31. McKenzie, L.W.: The Dorfman–Samuelson–Solow turnpike theorem. Int. Econ. Rev. **4**, 29–43 (1963)
32. McKenzie, L.W.: The turnpike theorem of Morishima. Rev. Econ. Stud. **30**, 169–176 (1963)
33. McKenzie, L.W.: Accumulation programs of maximum utility and the von Neumann facet. In: Wolfe, J.N. (ed.) Value, Capital and Growth, pp. 353–383. Edinburgh University Press, Edinburgh (1968)
34. McKenzie, L.W.: Turnpike theory. Econometrica **43**, 841–865 (1976)
35. McKenzie, L.W.: A new route to the turnpike. In: Henn, R., Moeschlin, L.O. (eds.) Mathemtical Economics and Game Theory. Springer, Berlin (1977)
36. McKenzie, L.W.: Turnpike theory, discounted utility, and the von Neumann facet. J. Econ. Theory **27**, 194–209 (1982)
37. McKenzie, L.W.: A primal route to the turnpike and Lyapunov stability. J. Econ. Theory **30**, 330–352 (1983)
38. McKenzie, L.W.: Optimal economic growth, turnpike theorems and comparative dynamics. In: Arrow, K.J., Intrilligator, M. (eds.) Handbook of Mathematical Economics, vol. 3, pp. 1281–1355. North-Holland, New York (1986)
39. McKenzie, L.W.: Turnpikes. Am. Econ. Rev. (Papers and Proceedings) **88**, 1–14 (1998)
40. McKenzie, L.W.: Classical General Equilibrium Theory. The MIT Press, Cambridge (2002)
41. Mitra, T.: Characterization of the turnpike property of optimal paths in the aggregative model of intertemporal allocation. Int. J. Econ. Theory **1**, 247–275 (2005)
42. Mitra, T., Wan, H.W., Jr.: Some theoretical results on the economics of forestry. Rev. Econ. Stud. **LII**, 263–282 (1985)
43. Mitra, T., Wan, H.W., Jr.: On the Faustmann solution to the forest management problem. J. Econ. Theory **40**, 229–249 (1986)
44. Pietsch, A.: History of Banach Spaces and Linear Operators. Birkhäuser, Berlin (2007)
45. Ramsey, F.: A mathematical theory of savings. Econ. J. **38**, 543–559 (1928)
46. Rockafellar, R.T.: Hamiltonian trajectories and saddle points in mathemtical economics. Contr. Cybern. **4**, 1575–1588 (2009)
47. Salo, S., Tahvonen, O.: On equilibrium cycles and normal forests in optimal harvesting of tree vintages. J. Environ. Econ. Manage. **44**, 1–22 (2002)

48. Salo, S., Tahvonen, O.: On the economics of forest vintages. J. Econ. Dyn. Control **27**, 1411–1435 (2003)
49. Samuelson, P.A.: A catenary turnpike theorem involving consumption and the golden rule. Am. Econ. Rev. **55**, 486–496 (1965). An erratum in 55, 864–866
50. Samuelson, P.A.: The general saddlepoint property of optimal control motions. J. Econ. Theory **5**, 102–120 (1972)
51. Samuelson, P.A., Solow, R.M.: A complete capital model involving heterogeneous capital goods. Q. J. Econ. **27**, 537–562 (1956)
52. Scheinkman, J.A.: On optimal steady states of n-sector growth models when utility is discounted. J. Econ. Theory **12**, 11–30 (1976)
53. Solow, R.M.: Modern capital theory. In: Brown, E.C., Solow, R.M. (eds.) Paul Samuelson and Modern Economic Theory. McGraw-Hill, New York (1983)
54. Stiglitz, J.E.: A note on technical choice under full employment in a socialist economy. Econ. J. **78**, 603–609 (1968)
55. von Neumann, J.: Über ein ökonomisches gleichungs-system und eine verallgemeinerung des Brouwerschen fixpunktsatzes. In: Menger, K. (ed.) Ergebnisse eines Mathematischen Kolloquiums, 8. Translated as: A model of general economic equilibrium. Rev. Econ. Stud. **13**, 1945–1946, 1–9 (1935–1936)
56. Yano, M.: Note on the existence of an optimal capital accumulation path. J. Econ. Theory **27**, 421–429 (1981)
57. Zaslavski, A.J.: Optimal programs in the RSS model. Int. J. Econ. Theory **1**, 151–165 (2005)
58. Zaslavski, A.J.: Turnpike Properties in the Calculus of Variations and Optimal Control. Springer, New York (2005)
59. Zaslavski, A.J.: Turnpike results for discrete-time optimal control systems arising in economic dynamics. Nonlinear Anal. **67**, 2024–2049 (2007)

Note added to proof: The Editor of *The New Palgrave* (2008), Professor Steven Durlauf, has very kindly informed the authors that the omission of the entry of the (late) Lionel McKenzie on *turnpike theory* was an oversight, and that it will be reinserted in the on-line version.

Adv. Math. Econ. 14, 69–92 (2011)

Advances in
MATHEMATICAL
ECONOMICS

Stock price process and long memory in trade signs

Koji Kuroda[1], **Jun-ichi Maskawa**[2], **and Joshin Murai**[3,*]

[1] Graduate school of Integrated Basic Sciences, Nihon University, Sakura-Josui 3-25-40, Setagaya-ku, Tokyo 156-8550, Japan
(e-mail: kuroda@math.chs.nihon-u.ac.jp)

[2] Department of Economics, Seijo University 6-1-20 Seijo, Setagaya-ku, Tokyo 157-8511, Japan
(e-mail: maskawa@gmail.com)

[3] Graduate school of Humanities and Social Sciences, Okayama University, Okayama 700-8530, Japan
(e-mail: murai@e.okayama-u.ac.jp)

Received: February 27, 2010
Revised: July 2, 2010

Mathematics Subject Classification (2010): 60K35, 60G22, 60F17, 82B20

Abstract. Empirical study on tick by tick data in stock markets shows us that there exists a long memory in trade signs and signed trade volumes. This means that an order flow is a highly autocorrelated long memory process.

We present a mathematical model of trade signs and trade volumes in which traders decompose their orders into small pieces. We prove that fractional Brownian motions are obtained as a scaling limit of the signed volume process induced by the model.

Key words: trade signs, long memory, Hurst index, fractional Brownian motion, Polymer expansion

1. Introduction

In recent years, a time series of trade signs has been investigated empirically by Bouchaud et al. [1] in the Paris Stock Exchange, and Lillo and Farmer [14] in the London Stock Exchange, and a long memory property was found

* Supported by Grant-in-Aid for Scientific Research (C) No. 21510146.

DOI: 10.1007/978-4-431-53883-7_4,

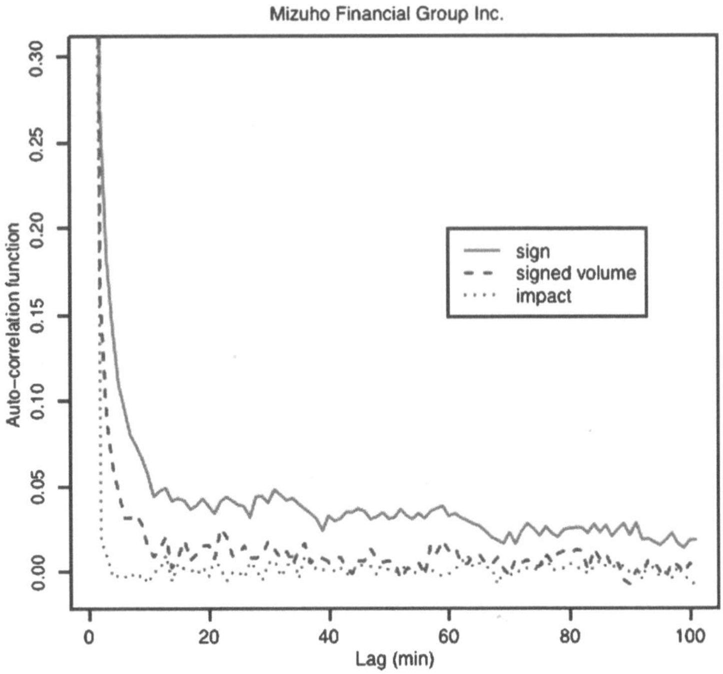

Fig. 1 Auto-correlation functions of of sequences of order signs, signed transaction volumes and price impacts per minute. The order sign and the signed volume are defined by the sing of (♯ of transactions executed at ask – ♯ of transactions executed at bid) per minute and (transaction volumes traded at ask – transaction volumes traded at bid) per minute respectively. We study the midday transactions of liquid securities. Here we show a result for Mizuho UFJ Financial Group Inc. for the period from Jan. to Dec. in 2008. The auto-correlation functions of order sign and signed volume decay slowly, showing power law tails, while the price impact has no significant auto-correlation

in the autocorrelation function of trade signs. It is known that stock price processes are well described by a Brownian motion having no long memory [1, 2]. However, an order flow is a highly autocorrelated long memory process. Bouchaud et al. [1] also showed that a long memory property exists in a time series of signed volumes. We also observed the same property in the Tokyo Stock Exchange [12]. (See also Fig. 1).

In many stock markets including the New York Stock Exchange, the London Stock Exchange and the Tokyo Stock Exchange *a continuous double auction system* is adopted as a standard mechanism of trade. If a buy(sell) market order arrives at time t it is matched with a sell(buy) limit order at the best ask(bid) price. We put a trade sign $s_t = +1(-1)$ on this buyer(seller)-initiated transaction. However, limit orders often fail to result in an immediate transaction and are stored in a queue called *limit order book*. In this article, we only consider the time series of trade signs and signed trade volumes for market orders.

A time series is said to have a long memory if its autocorrelation function $\rho(t)$ decays asymptotically as

$$\rho(t) \sim \frac{c}{t^{\alpha}} \qquad (t \to \infty)$$

where $\rho(t)$ is defined by

$$\rho(t) = \frac{\mathrm{Cov}(s_u, s_{u+t})}{\sqrt{V[s_u]V[s_{u+t}]}}$$

as a function of time lag t, c is some positive constant and $0 < \alpha < 1$.

Note that an integral of autocorrelation function with a long memory property diverges, because $0 < \alpha < 1$.

Also, a long memory process X_t can be discussed in terms of Hurst index H intuitively defined by $(dX_t) \propto (dt)^{2H}$. The fractional Brownian motion B_t^H with Hurst index H is known as a long memory process. It is defined as a Gaussian process having covariance,

$$\mathrm{Cov}(B_t^H, B_s^H) = \frac{1}{2}\left(t^{2H} + s^{2H} - |t-s|^{2H}\right).$$

When $H = \frac{1}{2}$, B_t^H is a standard Brownian motion. For a fractional Brownian motion B_t^H with the Hurst index $\frac{1}{2} < H < 1$, the autocorrelation function $\rho(\ell)$ of its increments $\Delta B_t^H = B_t^H - B_{t-1}^H$ is given by

$$\rho(\ell) = \frac{1}{2}\left[(\ell+1)^{2H} + (\ell-1)^{2H} - 2\ell^{2H}\right] \sim \frac{H(2H-1)}{\ell^{2-2H}} \qquad (\ell \to \infty), \tag{1}$$

hence it follows a power law relation with an exponent $0 < \alpha = 2-2H < 1$.

There are various explanations of the long memory property of trade signs. Among them, the dominant explanation is as follows (See [3]). Suppose that a trader decided to buy a large quantity of shares. If he reveals his intention to other traders, that brings upturn in stock price, and he needs to pay excess acquisition cost. It seems natural for him to keep his intention as secret as possible, to split them into small pieces and trades incrementally. Namely, the long memory is derived by the strategic behavior of large investors who split their orders into small pieces and execute them incrementally.

Lillo, Mike and Farmer obtain the discrete time stochastic process having a long memory property in the trade signs [15], assuming that the size of hidden orders is distributed as a power law and the time interval of placing orders is distributed uniformly.

In this article, we present a mathematical model of trade signs s_t and trade volumes v_t in discrete time interval $\Lambda_n = \{1, 2, \cdots, n\}$. We assume that traders can split their orders into m pieces, placed at $u_1, \cdots, u_m$ and that time intervals of placing orders are distributed as a power law,

$$\frac{1}{(u_2 - u_1)^\alpha \cdots (u_m - u_{m-1})^\alpha}$$

with exponent $0 < \alpha < 1$. Trade volumes $\{v_{u_1}, \cdots, v_{u_m}\}$ are assumed to be independent.

We consider a process W_ℓ of signed volumes defined by

$$W_\ell = \sum_{u=1}^{\ell} s_u v_u, \qquad (\ell \in \Lambda_n).$$

and a scale process

$$X_t^{(n)} = \frac{W_{[nt]}}{n^{\frac{1}{2}\alpha}} \qquad (0 < t < 1)$$

where $s_u \in \{+1, -1\}$ is a trade sign, v_u is a trade volume, $[x]$ is an integral part of $x \in \mathbf{R}$ and $0 < \alpha < 1$.

We derive superpositions of fractional Brownian motions with Hurst indexes $\{H_m\}$ as a limit of the scaled process $X_t^{(n)}$. The Hurst index H_m is determined by the splitting number m. In this model we do not consider a stock price process.

Introducing a notion of a *polymer* $\mathbf{p}$, we describe a set of trades by one trader as $\mathbf{p} = (s(\mathbf{p}), d(\mathbf{p}), b(\mathbf{p}), v(\mathbf{p}))$, where $s(\mathbf{p})$ is a trade sign, $d(\mathbf{p})$ is a number into which a trader split their order, $b(\mathbf{p})$ is a set of times in which orders are placed, and $v(\mathbf{p})$ is a set of trade volumes in $b(\mathbf{p})$.

As a mathematical tool for analyzing our model we use the method of abstract polymer expansion [5, 7, 9, 16] which has been investigated for the study of phase transitions in lattice spin systems. This method were applied to various problems in statistical physics such as the general theory of phase transition [4] and the study of interface between two phases [6, 8]. We also applied this method for the study of stock price process [10, 11, 13].

2. Description of the model and statement of result

Taking into account the idea by Lillo and Farmer [14] that the long memory property results from trader's behavior to split their orders into small pieces, we consider a discrete time stochastic model for trade signs and trade volumes of market orders.

Consider one security and assume that traders can place their market orders at any time $t \in \mathbf{Z}$. Furthermore, assume that two traders cannot place their orders at the same time, and that any market order placed by a trader can be transacted immediately.

We setup a model for trade signs and trade volumes on $\mathbf{Z}$ and observe the transactions in $\Lambda_n = \{1, \cdots, n\} \subset \mathbf{Z}$, to define a process W_ℓ, $(\ell \in \Lambda_n)$ of accumulated signed volumes.

To describe transactions of market orders split by a trader we introduce a notion of *polymer* defined by

$$\mathbf{p} = (s, m, \mathbf{u}, \mathbf{v}) \in \{+1, -1\} \times \{1, \cdots, m_0\} \times T_m^{(n)} \times V^m$$

where $m_0 \geq 2$ is a positive integer,

$$T_m^{(n)} = \{(u_1, \cdots, u_m) \in \mathbf{Z}^m ; 0 < u_i - u_{i-1} \leq n \;\; (i = 2, \cdots, m \;) \}$$

and $V = \{v^1, \cdots, v^M\}$ is a finite set of positive real numbers and we consider a probability distribution $P(\cdot)$ on V.

When $\mathbf{p} = (+1(-1), m, u_1, \cdots, u_m, v_{u_1}, \cdots, v_{u_m})$, this means that a trader split his buy(sell) market order into m pieces and place them at $u_1, \cdots, u_m \in \mathbf{Z}$ with trade volumes $v_{u_1}, \cdots, v_{u_m}$. Let $\mathcal{P}_n$ be a set of all polymers on $\mathbf{Z}$. Remark that one trade sign is assigned to $\mathbf{p}$, which means that a trader does not change his market order from buy(sell) position to sell(buy) position in $\mathbf{p}$.

For any $\mathbf{p} = (s, m, u_1, \cdots, u_m, v_{u_1}, \cdots, v_{u_m}) \in \mathcal{P}_n$ we put

$$s(\mathbf{p}) = s, \; |\mathbf{p}| = m, b(\mathbf{p}) = \{u_1, \cdots, u_m\} \subset \mathbf{Z}.$$

We call a set of polymers $\{\mathbf{p}_1, \cdots, \mathbf{p}_k\} \subset \mathcal{P}_n$ a *compatible polymer configuration* if

$$b(\mathbf{p}_i) \cap b(\mathbf{p}_j) = \emptyset, \; (1 \leq i < j \leq k).$$

A space Ω_n of compatible polymer configurations having intersection with Λ_n is defined by

$$\begin{aligned} \Omega_n = \{ \; \omega = \{\mathbf{p}_1, \cdots, \mathbf{p}_k\} &: \text{compatible polymer configuration;} \\ & k \in \mathbf{N}, \; b(\mathbf{p}_i) \cap \Lambda_n \neq \emptyset \; (i = 1, \cdots, k) \; \}. \end{aligned}$$

Note that Ω_n is a finite set for each $n \in \mathbf{N}$.

To define a probability distribution on Ω_n, we introduce a probability intensity function $\varphi(\mathbf{p})$ for $\mathbf{p} = (s, m, u_1, \cdots, u_m, v_{u_1}, \cdots, v_{u_m})$ defined by

$$\varphi(\mathbf{p}) \;=\; \begin{cases} d_n \cdot P(v_{u_1}) & (|\mathbf{p}| = 1) \\ d_n^m \frac{1}{(u_2-u_1)^\alpha \cdots (u_m-u_{m-1})^\alpha} \prod_{i=1}^m P(v_{u_i}) & (|\mathbf{p}| \geq 2) \end{cases} \tag{2}$$

where $0 < \alpha < 1, d_n = \frac{c}{n^{1-\alpha}} \quad (c > 0)$.

Let μ be an expectation value of trade volume given by

$$\mu = \sum_{i=1}^{M} v^i P(v^i).$$

A probability distribution of $\omega = \{\mathbf{p}_1, \cdots, \mathbf{p}_k\} \in \Omega_n$ is defined by

$$P_n(\omega) = \frac{1}{Z_n} \prod_{i=1}^{k} \varphi(\mathbf{p}_i) \tag{3}$$

where Z_n is a normalization constant.

For each $\omega = \{\mathbf{p}_1, \cdots, \mathbf{p}_k\} \in \Omega_n$ we define W_ℓ by

$$W_\ell(\omega) = \sum_{i=1}^{k} s(\mathbf{p}_i) \sum_{u=1}^{\ell} v_u(\mathbf{p}_i), \quad (\ell \in \Lambda_n)$$

and a scale process by

$$X_t^{(n)}(\omega) = \frac{W_{[nt]}(\omega)}{n^{\frac{1}{2}\alpha}} \quad (0 < t < 1)$$

where

$$v_u(\mathbf{p}) = \begin{cases} v_u & (u \in b(\mathbf{p})) \\ 0 & (\text{otherwise}), \end{cases}$$

and $[x]$ is an integral part of $x \in \mathbf{R}$.

In our model we only consider a process of accumulated signed volumes and do not consider a price process.

Theorem 2.1. *Assume that* $\frac{m_0 - 2}{m_0 - 1} < \alpha < 1$, *a finite dimensional distribution of* $X_t^{(n)}$ $(0 < t < 1)$ *converges to the corresponding distribution of*

$$\sqrt{b_0} B_t + \sum_{m=0}^{m_0-2} \sqrt{a_m} B_t^{H_m} \qquad (0 < t < 1)$$

where B_t *is a standard Brownian motion,* $B_t^{H_m}$ *is a fractional Brownian motion with Hurst index* $H_m = \frac{1}{2}((1-\alpha)(m+1)+1) \in \left(\frac{1}{2}, 1\right)$,

$$b_0 = 2\mu^2 c \left(1 + \sum_{m=2}^{m_0} m \left(\frac{c}{1-\alpha}\right)^{m-1}\right),$$

$$a_m = \frac{2E_m J_m c^{m+2}\mu^2}{(1-\alpha)(m+1)((1-\alpha)(m+1)+1)}, \; E_m = \sum_{k=0}^{m_0-(m+2)} \left(\frac{2c}{1-\alpha}\right)^k,$$

$$J_m = \iint_{\substack{0<x_1,\cdots,x_m<1\\ x_1+\cdots+x_m<1}} \frac{dx_1\cdots dx_m}{x_1^\alpha\cdots x_m^\alpha(1-x_1-\cdots-x_m)^\alpha}$$

and $B_t, B_t^{H_0}, \cdots, B_t^{H_{m_0-2}}$ *are independent.*

Since the Hurst index ought to be less than 1, we have assumed that $\frac{m_0-2}{m_0-1} < \alpha$ in order to fulfill the condition of $H_{m_0-2} < 1$.

3. Abstract polymer expansion

In this section we summarize a method of abstract polymer expansion which has been used for mathematical studies of statistical mechanics such as the general theory of phase transitions and the probabilistic approach to interfaces between two phases.

An abstract polymer expansion is a mathematical generalization of high temperature expansion or low temperature expansion in statistical mechanics. This method has been investigated by Gallavotti [7], Del Grosso [5], Pfister [16] and Kotecký-Preiss [9] from mathematical point of view and applied to various problems of phase transitions [4, 6, 8].

Only mutually disjoint polymer configuration is obtained in our model. But, we extend polymer configurations to more general configurations permitting multiplicities and intersections. Let $\mathcal{A}_n$ be a set of functionals A from $\mathcal{P}$ to $\mathbf{N} \cup \{0\}$,

$$\mathcal{A}_n = \left\{ A : \mathcal{P} \to \mathbf{N} \cup \{0\}; \quad |A| = \sum_{\mathbf{p}\in\mathcal{P}_n} A(\mathbf{p}) < \infty \right\}.$$

Each element A is regarded as a configuration of a finite number of polymers with multiplicity. We do not exclude the case that $b(\mathbf{p}_i) \cap b(\mathbf{p}_j) \neq \emptyset$.

For A_1 and $A_2 \in \mathcal{A}_n$ a sum $A_1 + A_2$ is defined by

$$(A_1 + A_2)(\mathbf{p}) = A_1(\mathbf{p}) + A_2(\mathbf{p}).$$

Also we define $A!$ for $A \in \mathcal{A}_n$ by $A! = \prod_{\mathbf{p}\in\mathcal{P}_n} A(\mathbf{p})!$.

Furthermore, we consider a functional space $\mathfrak{L}$ on $\mathcal{A}_n$,

$$\mathfrak{L} = \{\psi : \mathcal{A}_n \to \mathbf{R}; \quad \sup_{|A|=n}|\psi(A)| < \infty \text{ for all } n \geq 1\}.$$

and define a product $\psi_1 * \psi_2$ of ψ_1 and ψ_2 as follows,

$$\psi_1 * \psi_2(A) = \sum_{(A_1,A_2)\in\mathcal{A}_n\times\mathcal{A}_n;A_1+A_2=A} \frac{A!}{A_1!A_2!}\psi_1(A_1)\psi_2(A_2).$$

where the sum runs over all ordered (A_1, A_2) such that $A_1 + A_2 = A$.

With this product $*$ $\mathfrak{L}$ becomes a commutative algebra with identity $\mathbf{1}$ defined by

$$\mathbf{1}(A) = \begin{cases} 1 & \text{if } A = \emptyset \\ 0 & \text{otherwise.} \end{cases}$$

It is easily seen that $(\psi_1 * \psi_2) * \psi_3 = \psi_1 * (\psi_2 * \psi_3)$.

We define an exponential mapping Exp: $\mathfrak{L}_0 = \{\psi \in \mathfrak{L}; \psi(\emptyset) = 0\} \to \mathfrak{L}_1 = \{\psi \in \mathfrak{L} : \psi(\emptyset) = 1\}$ by

$$\mathrm{Exp}\psi(A) = \mathbf{1}(A) + \sum_{k=1}^{\infty}\frac{1}{k!}\underbrace{\psi * \cdots * \psi}_{k}(A).$$

As an inverse mapping of Exp a mapping Log: $\mathfrak{L}_1 \to \mathfrak{L}_0$ is defined by

$$\mathrm{Log}\psi(A) = \sum_{k=1}^{\infty}\frac{(-1)^{k+1}}{k}\underbrace{\psi_0 * \cdots * \psi_0}_{k}(A),$$

where $\psi_0 = \psi - \mathbf{1}$.

We call $\chi(\cdot) \in \mathfrak{L}$ *multiplicative* if $\chi(A_1 + A_2) = \chi(A_1)\chi(A_2)$.

The following lemma is a fundamental lemma for polymer expansion.

Lemma 3.1. ([5, 16]) *Let ψ be a functional in $\mathfrak{L}_0$ and χ be multiplicative. Assume that*

$$\sum_{A\in\mathcal{A}_n}|\chi(A)|\frac{|\psi^T(A)|}{A!} < \infty$$

then we have

$$\sum_{A\in\mathcal{A}_n}|\chi(A)|\frac{|\psi(A)|}{A!} < \infty \quad \textit{and}$$

$$\sum_{A\in\mathcal{A}_n}\chi(A)\frac{\psi(A)}{A!} = \exp\left\{\sum_{A\in\mathcal{A}_n}\chi(A)\frac{\psi^T(A)}{A!}\right\}$$

where $\psi^T(A) = \mathrm{Log}\psi(A)$.

To describe a probability distribution defined in (3) we introduce functionals;

$$\varphi(A) = \prod_{\mathbf{p}\in\mathcal{P}_n} \varphi(\mathbf{p})^{A(\mathbf{p})}$$

and $$\alpha(A) = \begin{cases} 1 & \text{if } A! = 1 \text{ and supp}A \text{ is a compatible polymer configuration} \\ 0 & \text{(otherwise)} \end{cases}$$

where $\text{supp}A = \{\mathbf{p} \in \mathcal{P}_n; A(\mathbf{p}) \geq 1\}$. .

In terms of these functionals we rewrite the probability distribution (3) as

$$P_n(A) = \begin{cases} \frac{1}{Z_n}\varphi(A)\alpha(A) & \text{if } A \cap \Lambda_n \neq \emptyset \\ 0 & \text{(otherwise)} \end{cases}$$

where Z_n is a normalization constant given by

$$Z_n = \sum_{A\in\mathcal{A}_n} \varphi(A)\alpha(A)$$

and $A \cap \Lambda_n \neq \emptyset$ means that $b(\mathbf{p}) \cap \Lambda_n \neq \emptyset$ for all $\mathbf{p} \in \text{supp}A$.

For any $A \in \mathcal{A}_n$ we assign a graph $\mathcal{G}(A)$. To each $\mathbf{p} \in \text{supp } A$ such that $A(\mathbf{p}) = k$ we assign k vertices of $\mathcal{G}(A)$ and draw edges between two vertices $\{\mathbf{p}_1, \mathbf{p}_2\}$ if $b(\mathbf{p}_1) \cap b(\mathbf{p}_2) \neq \emptyset$. The number of vertices is given by $|A| = \sum_{\mathbf{p}\in\mathcal{P}_n} A(\mathbf{p}) < \infty$. Employing this graph $\mathcal{G}(A)$ we have a graph description of $\alpha^T(A) = Log\alpha(A)$,

$$\alpha^T(A) = \sum_{C\subset\mathcal{G}(A):\text{connected}} (-1)^{|e(C)|}$$

where the sum runs over all connected subgraphs and $|e(C)|$ is a number of edges in C (see Sect. 3 of [16]). We say C is connected if the vertex set of C coincides with the vertex set of $\mathcal{G}(A)$ and all distinct vertices of C can be connected through edges of C.

We say A is a *cluster* if for any $\mathbf{p}^1 \in \text{supp } A$ and $\mathbf{p}^2 \in \text{supp } A$ there exists a chain $\mathbf{p}_0, \cdots, \mathbf{p}_m \in \text{supp } A$ of polymers such that

$$\mathbf{p}^1 = \mathbf{p}_0, \mathbf{p}^2 = \mathbf{p}_m, \text{ and } b(\mathbf{p}_k) \cap b(\mathbf{p}_{k+1}) \neq \emptyset \ (k = 0, \cdots, m-1).$$

In other words, cluster is a chain of connected polymers.

Then, we have an important property that $\alpha^T(A) = 0$ unless A is a cluster.

It is seen that $\varphi(A)$ and $\chi_n(A)$ are multiplicative , where $\chi_n(A)$ is defined by

$$\chi_n(A) = \begin{cases} 1 & (\text{if } A \cap \Lambda_n \neq \emptyset\ ) \\ 0 & (\text{otherwise}). \end{cases}$$

To apply Lemma 3.1 to our model we need to prepare some Lemmas. From standard argument of calculus we have the following lemma.

Lemma 3.2. *There exists a positive constant $C > 0$ such that the following estimate holds for sufficiently large n.*

$$\left| \frac{1}{n^m} \sum_{\substack{1 \le k_1, \cdots, k_m < n \\ k_1 + \cdots + k_m < n}} \frac{1}{(\frac{k_1}{n})^\alpha \cdots (\frac{k_m}{n})^\alpha (1 - \frac{k_1}{n} - \cdots - \frac{k_m}{n})^\alpha} - J_m \right| < \frac{C}{n^{1-\alpha}}.$$

Using this lemma for $m = 1$ we have the following lemma.

Lemma 3.3.

$$\sum_{\mathbf{p} \ni 0} e^{|\mathbf{p}|} \varphi(\mathbf{p}) = C_1 d_n + o(d_n)$$

where the sum runs over all $\mathbf{p}$ *with* $0 \in b(\mathbf{p})$ *and*

$$C_1 = 2 \left(e + \sum_{m=2}^{m_0} m e^m \left(\frac{c}{1-\alpha} \right)^{m-1} \right).$$

Decomposing connected graphs in $\alpha^T(A)$ into trees and estimating the number of trees, we obtain the following result from Lemma 3.3.

Lemma 3.4.

$$\sum_{A \ni 0} \varphi(A) \frac{\alpha^T(A)}{A!} = C_2 d_n + o(d_n)$$

where the sum runs over all A *with* $\bigcup_{\mathbf{p} \in \mathrm{supp} A} b(\mathbf{p}) \ni 0$ *and*

$$C_2 = 2 \left(1 + \sum_{m=2}^{m_0} m \left(\frac{c}{1-\alpha} \right)^{m-1} \right).$$

Proof.

We decompose the left hand side into two parts;

$$\sum_{A \ni 0} \varphi(A) \frac{\alpha^T(A)}{A!} = \sum_{\mathbf{p} \ni 0} \varphi(\mathbf{p}) + \sum_{\substack{A \ni 0 \\ |A| \ge 2}} \varphi(A) \frac{\alpha^T(A)}{A!}. \tag{4}$$

In the same way as Lemma 3.3 we have

$$\sum_{\mathbf{p}\ni 0} \varphi(\mathbf{p}) = C_2 d_n + o(d_n)$$

where

$$C_2 = 2\left(1 + \sum_{m=2}^{m_0-1} m\left(\frac{c}{1-\alpha}\right)^{m-1}\right).$$

Let us recall that the functional $\alpha^T(A)$ was described by a sum for connected subgraphs $C \subset \mathcal{G}(A)$. However, it is possible to describe $\alpha^T(A)$ as a sum for trees $\mathcal{T}$ satisfying some conditions. (See [16] for detail.)

To estimate the functional $\alpha^T(A)$ by the sum for trees, we prepare some terminologies. Let $\mathcal{G}(1, \cdots, n)$ be a set of all graphs having $\{1, \cdots, n\}$ as a vertex set. For any tree $\mathcal{T} \in \mathcal{G}(1, \cdots, n)$ we denote a set of all vertices of $\mathcal{T}$ and a set of all edges of $\mathcal{T}$ by $\mathcal{V}(\mathcal{T})$ and $\mathcal{E}(\mathcal{T})$, respectively.

As the functional $\alpha^T(A)$ is described by a tree satisfying some condition, we have

$$\sum_{\substack{A\ni 0;\\ |A|\geq 2}} \varphi(A)\frac{|\alpha^T(A)|}{A!} = \sum_{n\geq 2}\frac{n}{n!}\sum_{\mathbf{p}_1\ni 0}\sum_{\mathbf{p}_2}\cdots\sum_{\mathbf{p}_n}|\alpha^T(\mathbf{p}_1, \cdots, \mathbf{p}_n)|\prod_{i=1}^{n}\varphi(\mathbf{p}_i)$$

$$\leq \sum_{n\geq 2}\frac{n}{n!}\sum_{\mathcal{T}\in\mathcal{G}(1,\cdots,n)}\sum_{\mathbf{p}_1\ni 0}\sum_{\mathbf{p}_2}\cdots\sum_{\mathbf{p}_n}\prod_{(i,j)\in\mathcal{E}(\mathcal{T})} g(\mathbf{p}_i, \mathbf{p}_j)\prod_{i=1}^{n}\varphi(\mathbf{p}_i)$$

where the sum for $\mathcal{T}$ runs over all trees $\mathcal{T}$ in $\mathcal{G}(1, \cdots, n)$ and $g(\mathbf{p}_i, \mathbf{p}_j)$ is given by

$$g(\mathbf{p}_i, \mathbf{p}_j) = \begin{cases} 1 & \text{if } b(\mathbf{p}_i) \cap b(\mathbf{p}_j) \neq \emptyset \\ 0 & \text{(otherwise).} \end{cases}$$

For each vertex k we denote a number of edges starting from k by $d(k)$, and call it *a degree of* k.

Let us consider a set V_1 of vertices,

$$V_1 = \{k \in \mathcal{V}(\mathcal{T}); k \neq 1 \text{ and } d(k) = 1\ \}.$$

V_1 is a set of extremal vertices. To each $k \in V_1$ we assign a vertex $q(k) \in \mathcal{V}(\mathcal{T})$ uniquely such that $(k, q(k)) \in \mathcal{E}(\mathcal{T})$ (See Fig. 2).

Let us denote that $\mathcal{V}(\mathcal{T}) \setminus V_1 = \{1, \cdots, m_1\}$. Since each $\mathbf{p}_k$ $(k \in V_1)$ intersects with $\mathbf{p}_{q(k)}$ we have

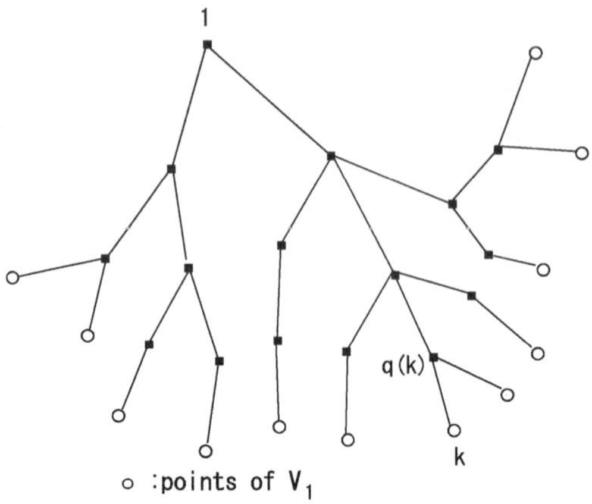

Fig. 2 Points of V_1

$$\sum_{\mathbf{p}_1 \ni 0} \sum_{\mathbf{p}_2} \cdots \sum_{\mathbf{p}_n} \prod_{(i,j)\in\mathcal{E}(\mathcal{T})} g(\mathbf{p}_i, \mathbf{p}_j) \prod_{i=1}^{n} \varphi(\mathbf{p}_i)$$
$$\leq \sum_{\mathbf{p}_1 \ni 0} \sum_{\mathbf{p}_2} \cdots \sum_{\mathbf{p}_{m_1}} \prod_{k\in V_1} \left(|\mathbf{p}_{q(k)}| \sum_{\mathbf{p}\ni 0} \varphi(\mathbf{p}) \right) \prod_{(i,j)\in\mathcal{E}(\mathcal{T}_1)} g(\mathbf{p}_i, \mathbf{p}_j) \prod_{i=1}^{m_1} \varphi(\mathbf{p}_i)$$

where $\mathcal{T}_1$ is a tree obtained from $\mathcal{T}$ by eliminating all vertices of V_1 and all edges starting from V_1.

In the same way as V_1 we consider a set V_2 of extremal vertices of $\mathcal{V}(\mathcal{T}_1)$ and we also denote that $\mathcal{V}(\mathcal{T}) \setminus (V_1 \cup V_2) = \{1, \cdots, m_2\}$, then we have

$$\sum_{\mathbf{p}_1 \ni 0} \sum_{\mathbf{p}_2} \cdots \sum_{\mathbf{p}_n} \prod_{(i,j)\in\mathcal{E}(\mathcal{T})} g(\mathbf{p}_i, \mathbf{p}_j) \prod_{i=1}^{n} \varphi(\mathbf{p}_i)$$
$$\leq \sum_{\mathbf{p}_1 \ni 0} \sum_{\mathbf{p}_2} \cdots \sum_{\mathbf{p}_{m_2}} \prod_{k\in V_1} \left(\sum_{\mathbf{p}\ni 0} \varphi(\mathbf{p}) \right) \prod_{k\in V_2} \left(\sum_{\mathbf{p}\ni 0} |\mathbf{p}_k|^{d(k)-1} \varphi(\mathbf{p}) \right)$$
$$\times \left(\prod_{k\in V_2} |\mathbf{p}_{q(k)}| \right) \prod_{(i,j)\in\mathcal{E}(\mathcal{T}_2)} g(\mathbf{p}_i, \mathbf{p}_j) \prod_{i=1}^{m_2} \varphi(\mathbf{p}_i)$$

where $\mathcal{T}_2$ is a tree obtained from $\mathcal{T}_1$ by eliminating all vertices of V_2 and all edges starting from V_2. In the above inequality we have used a relation

$$\prod_{k\in V_1} |\mathbf{p}_{q(k)}| = \prod_{k\in V_2} |\mathbf{p}_k|^{d(k)-1}.$$

Repeating these treatments we have

$$\sum_{\mathbf{p}_1 \ni 0} \sum_{\mathbf{p}_2} \cdots \sum_{\mathbf{p}_n} |\alpha^T(\mathbf{p}_1, \cdots, \mathbf{p}_n)| \prod_{i=1}^{n} \varphi(\mathbf{p}_i)$$
$$\leq \sum_{\mathcal{T} \in \mathcal{G}(1,\cdots,n):\ \text{tree}} \sum_{\mathbf{p}_1 \ni 0} |\mathbf{p}_1|^{d(1)} \varphi(\mathbf{p}_1) \prod_{k=2}^{n} \left(\sum_{\mathbf{p} \ni 0} |\mathbf{p}|^{d(k)-1} \varphi(\mathbf{p}) \right),$$

where $\{d(1), \cdots, d(n)\}$ is a set of degrees of points $\{1, \cdots, n\}$ for tree $\mathcal{T}$.

When a set of degrees of n vertices is given by $\{d(1), \cdots, d(n)\}$, the number of such trees is equal to

$$\binom{n-2}{d(1)-1, \cdots, d(n)-1} = \frac{(n-2)!}{(d(1)-1)! \cdots (d(n)-1)!}.$$

Hence, we have

$$\sum_{A \ni 0;\, |A| \geq 2} \varphi(A) \frac{|\alpha^T(A)|}{A!} \leq \sum_{n=2}^{\infty} \left(\sum_{\mathbf{p} \ni 0} \sum_{d=0}^{\infty} \frac{|\mathbf{p}|^d}{d!} \varphi(\mathbf{p}) \right)^n$$
$$= \sum_{n=2}^{\infty} \left(\sum_{\mathbf{p} \ni 0} e^{|\mathbf{p}|} \varphi(\mathbf{p}) \right)^n = O(d_n^2) \qquad (n \to \infty).$$

q.e.d.

From this Lemma it follows that

$$\sum_{A \cap \Lambda_n \neq \emptyset} \varphi(A) \frac{|\alpha^T(A)|}{A!} \leq C n^{\alpha} < \infty$$

for some $C > 0$.

Now we can apply Lemma 3.1 to $\varphi(A)\chi_n(A)\alpha(A)$ and obtain

$$Z_n = \exp \left\{ \sum_{A \cap \Lambda_n \neq \emptyset} \varphi(A) \frac{\alpha^T(A)}{A!} \right\}.$$

4. Proof of result

We consider a signed volume process,

$$W_\ell(A) = \sum_{\mathbf{p}} A(\mathbf{p}) s(\mathbf{p}) \sum_{u=1}^{\ell} v_u(\mathbf{p}), \quad (\ell \in \Lambda_n)$$

and scale process $X_t^{(n)}(A)$ defined by

$$X_t^{(n)}(A) = \frac{W_{[nt]}(A)}{n^{\frac{1}{2}\alpha}}, \quad (0 < t < 1).$$

Proof of the Theorem 2.1

We consider a characteristic function $\varphi_{t_1,\cdots,t_r}(z_1, \cdots, z_r)$ of $(X_{t_1}^{(n)}, \cdots, X_{t_r}^{(n)})$ $(0 < t_1 < \cdots < t_r < 1)$ given by

$$\varphi_{t_1,\cdots,t_r}(z_1, \cdots, z_r) = E_n \left[\exp\left\{ i \sum_{k=1}^{r} z_k X_{t_k}^{(n)} \right\} \right].$$

To prove the theorem we have only to show that

$$\varphi_{t_1,\cdots,t_r}(z_1, \cdots, z_r) \to \exp\left\{ -\frac{1}{2} \sum_{k=1}^{r} \sum_{\ell=1}^{r} z_k z_\ell \left(b_0(t_k \wedge t_\ell) + \sum_{m=0}^{m_0-2} \frac{a_m}{2} (t_k^{2H_m} + t_\ell^{2H_m} - |t_k - t_\ell|^{2H_m}) \right) \right\} \quad (n \to \infty).$$

Employing Lemma 3.4, we can apply Lemma 3.1 to $\varphi_{t_1,\cdots,t_r}(z_1, \cdots, z_r)$ and get

$$\begin{aligned}
\log \varphi_{t_1,\cdots,t_r}(z_1, \cdots, z_r) &= \frac{i}{c(n)} \sum_{k=1}^{r} z_k \sum_{A \cap \Lambda_n \neq \emptyset} W_{[nt_k]}(A) \varphi(A) \frac{\alpha^T(A)}{A!} \\
&- \frac{1}{2c(n)^2} \sum_{k=1}^{r} \sum_{\ell=1}^{r} z_k z_\ell \sum_{A \cap \Lambda_n \neq \emptyset} W_{[nt_k]}(A) W_{[nt_\ell]}(A) \varphi(A) \frac{\alpha^T(A)}{A!} \\
&- \frac{i}{6c(n)^3} \sum_{A \cap \Lambda_n \neq \emptyset} \left(\sum_{k=1}^{r} z_k W_{[nt_k]}(A) \right)^3 \\
&\times \exp\left\{ i\theta \frac{1}{c(n)} \sum_{k=1}^{r} z_k W_{[nt_k]}(A) \right\} \varphi(A) \frac{\alpha^T(A)}{A!} \qquad (5)
\end{aligned}$$

for some $\theta \in (0, 1)$.

It follows from $\pm$ symmetry the first term of (5) becomes zero.

We shall prove that the second term of (5) converges to the covariance terms of fractional Brownian motions.

To describe the second term more explicitly we introduce a functional $Z_k(A)$ defined by

$$Z_k(A) = \sum_{\mathbf{p};k\in b(\mathbf{p})} A(\mathbf{p})s(\mathbf{p})v_k(\mathbf{p})$$

and rewrite

$$I_{k,\ell} := \sum_{A\cap\Lambda_n\neq\emptyset} W_{[nt_k]}(A)W_{[nt_\ell]}(A)\varphi(A)\frac{\alpha^T(A)}{A!}$$

as

$$I_{k,\ell} = \sum_{j=1}^{[nt_k]}\sum_{p=1}^{[nt_\ell]}\sum_{A\cap\Lambda_n\neq\emptyset} Z_j(A)Z_p(A)\varphi(A)\frac{\alpha^T(A)}{A!}.$$

Lemma 4.2.

For any $\xi \in (0, 1)$, we have the following relation if $w = |j - p| > n^\xi$ for sufficiently large n,

$$\begin{aligned}\hat{I}_{j,p} &:= \sum_{A\cap\Lambda_n\neq\emptyset} Z_j(A)Z_p(A)\varphi(A)\frac{\alpha^T(A)}{A!}\\ &= \sum_{m=0}^{m_0-2} I_m(w)\cdot\left(1 + O\left(\frac{1}{n^{\xi(1-\alpha)}}\right)\right)\end{aligned}$$

where

$$\begin{aligned}I_m(w) &= \frac{2E_mJ_md_n^{m+2}\mu^2}{w^{\alpha-(1-\alpha)m}}\\ J_0 &= 1\\ J_m &= \int\int_{\substack{0<x_1,\cdots,x_m<1\\ x_1+\cdots+x_m<1}} \frac{1}{x_1^\alpha x_2^\alpha\cdots x_m^\alpha(1-x_1-\cdots-x_m)^\alpha}dx_1\cdots dx_m\\ &\quad\times(1\le m\le m_0-2)\\ E_m &= \sum_{k=0}^{m_0-(m+2)}\left(\frac{2c}{1-\alpha}\right)^k.\end{aligned}$$

Proof.

Suppose that $j < p$. Contribution from terms for A in which a polymer $\mathbf{p}_1 \in \operatorname{supp} A$ containing j is different from a polymer $\mathbf{p}_2 \in \operatorname{supp} A$ containing p becomes zero from $\pm$ symmetry. So that we only consider terms in which a polymer $\mathbf{p}_0 \in \operatorname{supp} A$ contains both j and p.

Then we can decompose $\hat{I}_{j,p}$ into two terms,

$$\hat{I}_{j,p} = \sum_{\substack{A;\exists \mathbf{p}_0 \in \operatorname{supp} A;\\ j,p \in b(\mathbf{p}_0),\, |A|=1}} Z_j(A) Z_p(A) \varphi(A) \frac{\alpha^T(A)}{A!} + \sum_{\substack{A;\exists \mathbf{p}_0 \in A;\\ j,p \in b(\mathbf{p}_0),\, |A|\geq 2}} Z_j(A) Z_p(A) \varphi(A) \frac{\alpha^T(A)}{A!}. \tag{6}$$

We consider the first term of (6) where A is composed of a single polymer $\mathbf{p}_0$.

First term of (6)

$$= \sum_{m=0}^{m_0-2} 2 d_n^{m+2} \mu^2 E_m^{(n)} \sum_{0<u_1<\cdots<u_m<w} \frac{1}{u_1^\alpha (u_2-u_1)^\alpha \cdots (u_m-u_{m-1})^\alpha (w-u_m)^\alpha}$$

$$= \sum_{m=0}^{m_0-2} \frac{2 d_n^{m+2} \mu^2 E_m^{(n)}}{w^{\alpha-(1-\alpha)m}} \cdot \frac{1}{w^m} \sum_{0<u_1<\cdots<u_m<w} \frac{1}{(\frac{u_1}{w})^\alpha (\frac{u_2}{w}-\frac{u_1}{w})^\alpha \cdots (\frac{u_m}{w}-\frac{u_{m-1}}{w})^\alpha (1-\frac{u_m}{w})^\alpha}$$

where

$$E_m^{(n)} = \sum_{k=0}^{m_0-(m+2)} 2^k \left(d_n \sum_{u=1}^{n} \frac{1}{u^\alpha} \right)^k \to E_m = \sum_{k=0}^{m_0-(m+2)} 2^k \left(\frac{c}{1-\alpha} \right)^k. \tag{7}$$

Remark that m is a number of points of $b(\mathbf{p}_0)$ between j and p, and E_m is a contribution from points on the left-hand side of j and points on the right-hand side of p .

From Lemma 3.2 we have

$$\text{The first term of (6)} = \sum_{m=0}^{m_0-2} I_m(w) \cdot \left(1 + O\left(\frac{1}{n^{\xi(1-\alpha)}}\right) \right).$$

The second term of (6) can be estimated by Lemma 3.4 and the proof is completed. (q.e.d.)

Now we shall apply Lemma 4.2 to $I_{k,\ell}$. Suppose that $t_\ell < t_k$. For simplicity we use the notation $a_n \sim b_n$ if $a_n/b_n \to 1$ as $n \to \infty$.

We decompose a domain of the sum for (j, p) into five regions $R_1, \sim, R_5$ given by

$$
\begin{aligned}
R_1 &= \{(j, p); 1 \le j = p \le [nt_\ell]\}, \\
R_2 &= \{(j, p); 0 < |j - p| \le n^\xi, 1 \le p \le [nt_\ell], 1 \le j\}, \\
R_3 &= \{(j, p); 1 \le j, j + n^\xi < p \le [nt_\ell]\}, \\
R_4 &= \{(j, p); 1 \le p \le [nt_\ell], j - [n(t_k - t_\ell)] < p \le j - n^\xi \}, \\
R_5 &= \{(j, p); 1 \le p \le j - [n(t_k - t_\ell)], [n(t_k - t_\ell)] \le j \le [nt_k] \},
\end{aligned}
$$

(See Fig. 3) and define J_k $(k = 1, \cdots, 5)$ by

$$J_k = \sum_{(j,p)\in R_k} \hat{I}_{j,p}.$$

First we consider a diagonal term J_1,

$$
\begin{aligned}
J_1 &\sim nt_\ell \sum_{A \ni 0} Z_0(A)^2 \varphi(A) \frac{\alpha^T(A)}{A!} \\
&\sim \mu^2 C_2 d_n \cdot nt_\ell \\
&= (\mu^2 C_2 c) \cdot t_\ell \cdot n^\alpha. \qquad (8)
\end{aligned}
$$

We shall prove that $J_2 = o(n^\alpha)$. From the standard argument we have

$$
\begin{aligned}
|J_2| &\le 2nt_\ell d_n^2 \cdot n^\xi \cdot \sum_{A \ni 0} \frac{\varphi(A)}{d_n} \cdot \frac{|\alpha^T(A)|}{A!} \\
&\le Cn^{2\alpha - 1 + \xi}
\end{aligned}
$$

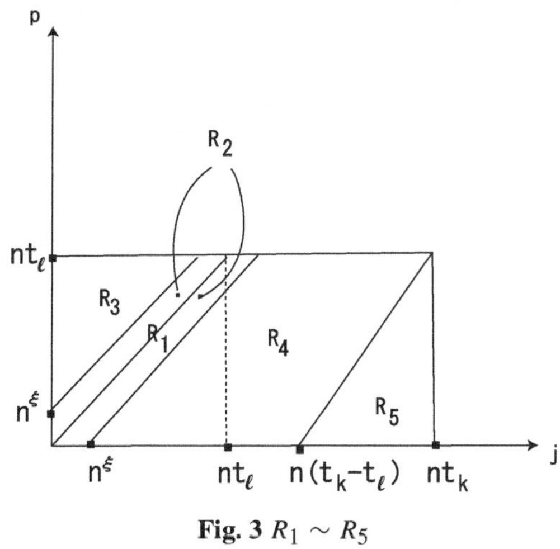

Fig. 3 $R_1 \sim R_5$

for some constant $C > 0$. Take a value of ξ satisfying $0 < \xi < 1 - \alpha$, then $J_2 = o(d_n)$.

Using Lemma 4.2 we decompose the main term of J_3

$$J_3 \sim \sum_{m=0}^{m_0-2} J_3^{(m)}$$

where

$$J_3^{(m)} = \sum_{w=1}^{[nt_\ell]} ([nt_\ell] - w) I_m(w).$$

Then we have,

$$\begin{aligned} J_3^{(m)} &\sim 2E_m J_m d_n^{m+2} \mu^2 \sum_{w=1}^{[nt_\ell]} ([nt_\ell] - w) \frac{1}{w^{\alpha-(1-\alpha)m}} \\ &\sim 2E_m J_m d_n^{m+2} \mu^2 \left(nt_\ell \int_1^{nt_\ell} w^{(1-\alpha)m-\alpha} dw - \int_1^{nt_\ell} w^{(1-\alpha)(m+1)} dw \right) \\ &= 2E_m J_m c^{m+2} \mu^2 \cdot \frac{t_\ell^{(1-\alpha)(m+1)+1}}{(1-\alpha)(m+1)((1-\alpha)(m+1)+1)} \cdot n^\alpha. \end{aligned}$$

We define $J_4^{(m)}$ and $J_5^{(m)}$ in the same way as $J_3^{(m)}$, then we have

$$\begin{aligned} J_4^{(m)} &\sim \frac{2E_m J_m c^{m+2} \mu^2}{(1-\alpha)(m+1)} \cdot t_\ell (t_k - t_\ell)^{(1-\alpha)(m+1)} \cdot n^\alpha \\ J_5^{(m)} &\sim 2E_m J_m c^{m+2} \mu^2 \left\{ \frac{t_k^{(1-\alpha)(m+1)+1} - t_k (t_k - t_\ell)^{(1-\alpha)(m+1)}}{(1-\alpha)(m+1)} \right. \\ &\qquad \left. - \frac{t_k^{(1-\alpha)(m+1)+1} - (t_k - t_\ell)^{(1-\alpha)(m+1)+1}}{(1-\alpha)(m+1)+1} \right\} \cdot n^\alpha. \end{aligned}$$

Hence, we have

$$\begin{aligned} & J_3^{(m)} + J_4^{(m)} + J_5^{(m)} \\ & \quad \sim 2E_m J_m c^{m+2} \mu^2 \left\{ \frac{t_k^{(1-\alpha)(m+1)+1} + t_\ell^{(1-\alpha)(m+1)+1} - |t_k - t_\ell|^{(1-\alpha)(m+1)+1}}{(1-\alpha)(m+1)((1-\alpha)(m+1)+1)} \right\} \cdot n^\alpha. \end{aligned}$$

Putting these estimates together we have the following proposition for the second term of $\varphi_{t_1,\cdots,t_r}(z_1, \cdots, z_r)$.

Proposition 4.3.

$$\lim_{n\to\infty}\frac{1}{n^{\alpha}}\sum_{k=1}^{r}\sum_{\ell=1}^{r} z_k z_\ell \sum_{A\cap\Lambda_n\neq\emptyset} W_{[nt_k]}(A)W_{[nt_\ell]}(A)\varphi(A)\frac{\alpha^T(A)}{A!}$$
$$=\sum_{k=1}^{r}\sum_{\ell=1}^{r} z_k z_\ell \left(b_0(t_k\wedge t_\ell)+\sum_{m=0}^{m_0-2} a_m\cdot\frac{1}{2}\left(t_k^{(1-\alpha)(m+1)+1}+t_\ell^{(1-\alpha)(m+1)+1}\right.\right.$$
$$\left.\left.-|t_k-t_\ell|^{(1-\alpha)(m+1)+1}\right)\right)$$

where

$$b_0=\mu^2 C_2 c$$
$$a_m=\frac{2E_m J_m c^{m+2}\mu^2}{(1-\alpha)(m+1)((1-\alpha)(m+1)+1)}.$$

To complete the proof of the Theorem, we have only to prove that the third term of (5) converges to 0 as $n\to\infty$.

As a first step, we have

|The third term of (5)|

$$\le\frac{1}{6n^{\frac{3}{2}\alpha}}\sum_{A\cap\Lambda_n\neq\emptyset}\left(\sum_{i=1}^{r} z_k W_{[nt_i]}(A)\right)^2\sum_{u=1}^{r}|z_j||W_{[nt_j]}(A)|\varphi(A)\frac{|\alpha^T(A)|}{A!}$$
$$\le\frac{1}{6n^{\frac{3}{2}\alpha}}\sum_{i,u,v=1}^{r} z_i z_u|z_v|\sum_{k=1}^{[nt_i]}\sum_{\ell=1}^{[nt_u]}\sum_{j=1}^{[nt_v]}\sum_{A\cap\Lambda_n\neq\emptyset} Z_k(A)Z_\ell(A)|Z_j(A)|\varphi(A)\frac{|\alpha^T(A)|}{A!}.$$

Lemma 4.4.

There exists $C_3>0$ and the following relations hold,

(1)
$$\sum_{A\cap\Lambda_n\neq\emptyset} Z_k(A)Z_\ell(A)|Z_j(A)|\varphi(A)\frac{|\alpha^T(A)|}{A!}\le\frac{C_3 d_n^3}{(\ell-k)^{\alpha}(j-\ell)^{\alpha}}\qquad\text{when } k<\ell<j$$

(2)
$$\sum_{A\cap\Lambda_n\neq\emptyset} Z_k(A)|Z_j(A)|Z_\ell(A)\varphi(A)\frac{|\alpha^T(A)|}{A!}\le\frac{C_3 d_n^3}{(j-k)^{\alpha}(\ell-j)^{\alpha}}\qquad\text{when } k<j<\ell$$

Proof of (1):

From $\pm$ symmetry we have only to consider a case that k and ℓ are contained in a polymer $\mathbf{p}_0$. Contribution from other terms in which a polymer containing k is different from a polymer containing ℓ becomes 0 from $\pm$ symmetry.

We decompose the left-hand side into two parts;

$$\sum_{A\cap\Lambda_n\neq\emptyset} Z_k(A)Z_\ell(A)|Z_j(A)|\varphi(A)\frac{|\alpha^T(A)|}{A!} = I_1 + I_2$$

where

$$I_1 = \sum_{\substack{\exists\mathbf{p}_0\in\mathrm{supp}A:|A|=1\\ k,\ell\in b(\mathbf{p}_0)}} Z_k(A)Z_\ell(A)|Z_j(A)|\varphi(A)\frac{|\alpha^T(A)|}{A!}$$

$$I_2 = \sum_{\substack{\exists\mathbf{p}_0\in\mathrm{supp}A:|A|\geq 2\\ k,\ell\in b(\mathbf{p}_0)}} Z_k(A)Z_\ell(A)|Z_j(A)|\varphi(A)\frac{|\alpha^T(A)|}{A!}.$$

As $\mathbf{p}_0$ contains k, ℓ, and j in I_1, in a similar way to the proof of Lemma 3.4, we have

$$|I_1| \leq \frac{Cd_n^3}{(\ell-k)^\alpha(j-k)^\alpha}$$

for some $C > 0$.

To obtain a bound for the second term I_2 we consider a case that there exists a chain of polymers $\{\mathbf{p}_1, \mathbf{p}_2, \cdots, \mathbf{p}_{q-1}\}$ satisfying

$$\begin{aligned} j = p_0 &\in b(\mathbf{p}_0)\\ p_1 &\in b(\mathbf{p}_1)\cap b(\mathbf{p}_2)\\ p_2 &\in b(\mathbf{p}_2)\cap b(\mathbf{p}_3)\cdots\cdots\\ p_{q-1} &\in b(\mathbf{p}_{q-1})\cap b(\mathbf{p}_0)\end{aligned}$$

(See Fig. 4). This chain of polymers connects j and k.

Note that p_{k-1} and $p_k \in b(\mathbf{p}_k)$, we define a vector $\mathbf{w}_k$ by

$$\mathbf{w}_k = (p_{k-1}, p_k)$$

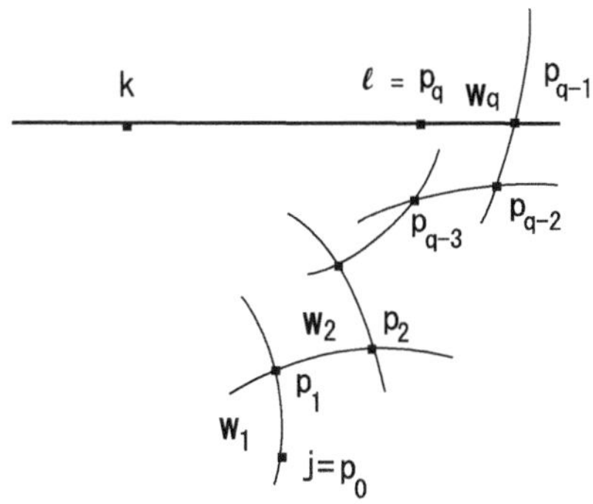

Fig. 4 Chain of polymers connecting ℓ and j

and define a probability intensity function $f(\mathbf{w}_k)$ of the occurrence of $\mathbf{w}_k$ by

$$f(\mathbf{w}_k) = 2\sum_{m=0}^{m_0-2} d_n^{m+2} E_m \sum_{\substack{0<u_1,\cdots,u_m \\ u_1+\cdots+u_m<\rho_k}} \frac{1}{u_1^{\alpha}\cdots u_m^{\alpha}(\rho_k - u_1 - \cdots - u_m)^{\alpha}}$$

where $\rho_k = |p_k - p_{k-1}|$.

When $\rho_k > n^{\xi}$, in a similar way to Lemma 3.4, we have

$$f(\mathbf{w}_k) \le C_4 \frac{d_n^2}{\rho_k^{\alpha}} \tag{9}$$

When $\rho_k \le n^{\xi}$, using an estimate

$$d_n \sum_{t=1}^{n^{\xi}} \frac{1}{t^{\alpha}} = O\left(\frac{1}{n^{(1-\xi)(1-\alpha)}}\right)$$

we have the same inequality as (9).

For $\mathbf{w} = (p_1, p_3)$ we define a convolution of $f(\cdot)$ by

$$f * f(\mathbf{w}) = \sum_{p_2 \in \mathbf{Z}} f(\mathbf{w}_1) f(\mathbf{w}_2)$$

where $\mathbf{w}_1 = (p_1, p_2)$, $\mathbf{w}_2 = (p_2, p_3)$ and the sum runs over all $p_2 \in \mathbf{Z}$.

From (9) and a standard argument of calculus it follows that

$$f * f(\mathbf{w}) \le K^2 \cdot d_n^3 \cdot \frac{1}{w^{\alpha}} \tag{10}$$

for some $K > 0$.

Repeating this process we have

$$\underbrace{f * \cdots * f}_{q}(\mathbf{w}) \leq K^q d_n^{q+1} \cdot \frac{1}{|j-\ell|^\alpha} \tag{11}$$

where $\mathbf{w} = (j, \ell)$.

Hence a probability that j and ℓ are connected by a chain of polymers is bounded by

$$d_n^2 \cdot \frac{1}{|j-\ell|^\alpha} \sum_{q=1}^{\infty} K^q d_n^{q-1} = d_n^2 \cdot \frac{1}{|j-\ell|^\alpha} \cdot \left(K + o\left(\frac{1}{n^{1-\alpha}}\right)\right).$$

Other cases are treated in a similar way. Hence we complete the proof of (1) in lemma 4.4.

A proof of (2) is also obtained in a similar way to (1). (q.e.d. of Lemma 4.4).

Now we shall prove that

$$\begin{aligned} I := \frac{1}{c(n)^3} \sum_{k=1}^{[nt_i]} \sum_{\ell=1}^{[nt_u]} \sum_{j=1}^{[nt_v]} \sum_{A \cap \Lambda_n \neq \emptyset} Z_k(A) Z_\ell(A) |Z_j(A)| \varphi(A) \frac{|\alpha^T(A)|}{A!} \\ \to 0 \qquad (n \to \infty). \end{aligned} \tag{12}$$

1° First we consider a case $k = \ell = j$ and consider I_1 given by

$$I_1 = \frac{1}{c(n)^3} \sum_{k=1}^{[nt]} |Z_k(A)|^3 \varphi(A) \frac{|\alpha^T(A)|}{A!}$$

where $t = \min\{t_i, t_u, t_v\}$.

It follows from Lemma 3.4 that

$$\begin{aligned} |I_1| &\leq K' \frac{1}{n^{\frac{3}{2}\alpha}} \cdot nt \cdot \mu^3 \cdot d_n \\ &= \frac{K' ct\mu^3}{n^{\frac{1}{2}\alpha}} \to 0 \qquad (n \to \infty) \end{aligned}$$

where $K' > 0$.

2° Consider a case $k = \ell \neq j$ and consider I_2 given by

$$I_2 = \frac{1}{c(n)^3} \sum_{k=1}^{[nt]} \sum_{j=1}^{[nt_v]} Z_k^2(A)|Z_j(A)|\varphi(A)\frac{|\alpha^T(A)|}{A!}$$

where $t = \min\{t_i, t_u\}$.

Using Lemma 4.2 we have

$$\begin{aligned} |I_2| &\leq \frac{\mu^3}{n^{\frac{3}{2}\alpha}} \sum_{w=1}^{[nt]} ([nt] - w)\frac{1}{w^\alpha} d_n^2 \\ &\leq K'' \cdot \frac{1}{n^{\frac{1}{2}\alpha}} \to 0 \qquad (n \to \infty) \end{aligned}$$

where $K'' > 0$.

3° Consider a case $k < \ell < j$, and consider I_3 given by

$$I_3 = \frac{1}{c(n)^3} \sum_{1 \leq k < \ell < j \leq [nt]} \sum_{A \cap \Lambda_n \neq \emptyset} Z_k(A) Z_\ell(A)|Z_j(A)|\varphi(A)\frac{|\alpha^T(A)|}{A!}$$

where $t = \max\{t_i, t_u, t_v\}$.

Using Lemma 4.4 we have

$$\begin{aligned} |I_3| &\leq \frac{C_3 d_n^3}{n^{\frac{3}{2}\alpha}} \sum_{1 \leq k < \ell < j \leq [nt]} \frac{1}{(\ell - k)^\alpha (j - \ell)^\alpha} \\ &\leq \frac{\text{constant}}{n^{\frac{3}{2}\alpha + 3(1-\alpha)}} \sum_{1 \leq k < j \leq [nt]} \frac{1}{|j - k|^{2\alpha - 1}} \\ &= \frac{\text{constant}}{n^{3 - \frac{3}{2}\alpha}} \sum_{w=1}^{[nt]} ([nt] - w)\frac{1}{w^{2\alpha - 1}} \\ &\leq \text{constant} \cdot \frac{1}{n^{\frac{1}{2}\alpha}} \to 0 \qquad (n \to \infty). \end{aligned}$$

Other cases are treated in a similar way, and the proof of (12) is obtained. Hence we complete the proof of Theorem 4.1.

References

1. Bouchaud, J.-P., Gefen, Y., Potters, M., Wyart, M.: Fluctuations and response in financial markets: the subtle nature of 'random' price changes. Quant. Finance **4**, 176–190 (2004)

2. Bouchaud, J.-P., Kockelkoren, J., Potters, M.: Random walks, liquidity molasses and critical response in financial markets. Quant. Finance **6**, 115–123 (2006)
3. Bouchaud, J.-P, Farmer, J.D., Lillo, F.: How markets slowly digest changes in supply and demand. In: Hens, T., Schenk-Hoppe, K. (eds.), Handbook of Financial Markets: Dynamics and Evolution, pp. 57–160 (2009)
4. Bricmont, J., Kuroda, K., Lebowitz, J.L.: First order phase transitions in lattice and continuous systems: extension of Pirogov–Sinai theory. Commun. Math. Phys. **101**, 501–538 (1985)
5. Del Grosso, G.: On the local central limit theorem for Gibbs processes. Commun. Math. Phys. **37**, 141–160 (1974)
6. Dobrushin, R.L., Hryniv, O.: Fluctuations of the phase boundary in the 2D Ising ferromagnet. Commun. Math. Phys. **189**, 395–445 (1997)
7. Gallavotti, G.: The phase separation line in the two-dimensional Ising model. Commun. Math. Phys. **27**, 103–136 (1972)
8. Higuchi, Y.: On some limit theorems related to the phase separation line in the two dimensional Ising model. Z. Wahrsch. Verv. Gebiete **50**, 287–315 (1979)
9. Kotecky, R., Preiss, D.: Cluster expansion for abstract polymer models. Commun. Math. Phys. **103**, 491–498 (1986)
10. Kuroda, K., Murai, J.: Stock price process and the long-range percolation. In: Takayasu, H. (ed.) Practical Fruits of Econophysics, pp. 163–167. Springer, Tokyo (2005)
11. Kuroda, K., Murai, J.: Limit theorems in financial market models. Physica A **383**, 28–34 (2007)
12. Kuroda, K., Murai, J.: A probabilistic model on the long memory property in stock market. In: International conference 2008 in Okayama, Rising Economies and Regional Cooperation in the East Asia and Europe, pp. 1–20 (2008)
13. Kuroda, K., Murai, J.: Long memory in finance and fractional Brownian motion. Progr. Theor. Phys. Suppl. **179**, 26–37 (2009)
14. Lillo, F., Farmer, J.: The Long Memory of the efficient market. Stud. Nonlinear Dyn. Econometrics **8**, 1–33 (2004)
15. Lillo, F., Mike, S., Farmer, J.D.: Theory for long memory in supply and demand. Phys. Rev. E **7106**(6 pt 2), 287–297 (2005)
16. Pfister, C.E.: Large deviations and phase separation in the two dimensional Ising model. Helv. Phys. Acta **64**, 953–1054 (1991)

Adv. Math. Econ. 14, 93–126 (2011)

Advances in MATHEMATICAL ECONOMICS

Extended second welfare theorem for nonconvex economies with infinite commodities and public goods

Aychiluhim Habte[1] **and Boris S. Mordukhovich**[2,3,*]

[1] Department of Mathematics and Computer Science, Benedict College, Columbia, SC 29204, USA
(e-mail: habtea@benedict.edu)

[2] Department of Mathematics, Wayne State University, Detroit, MI 48202, USA
(e-mail: aa1086@wayne.edu)

[3] Research of this author was partly supported by the US National Science Foundation under grant DMS-0603846, DMS-1007132 and by Australian Research Council under grant DP-12092508.

Received: April 12, 2010
Revised: June 28, 2010

JEL classification: D06

Mathematics Subject Classification (2010): 90A14, 49J52

Abstract. This paper is devoted to the study of nonconvex models of welfare economics with public goods and infinite-dimensional commodity spaces. Our main attention is paid to new extensions of the fundamental second welfare theorem to the models under consideration. Based on advanced tools of variational analysis and generalized differentiation, we establish appropriate approximate and exact versions of the extended second welfare theorem for Pareto, weak Pareto, and strong Pareto optimal allocations in both marginal price and decentralized price forms.

Key words: welfare economics, public goods, Pareto optimal allocations, second welfare theorem, variational analysis, generalized differentiation, extremal principle

* The second author gratefully acknowledges numerous discussions with Ali Khan on the subject of this paper and other issues of general equilibrium theory. We also thanks the referee for his/her careful reading the paper and helpful remarks.

DOI: 10.1007/978-4-431-53883-7_5,

1. Introduction

In this paper we consider general nonconvex models of welfare economics involving both private and public goods in infinite-dimensional spaces of commodities. Models of this type have been well recognized from both theoretical and practical viewpoints. Despite a number of excellent works in this area, there are great many of important unsolved problems some of which are addressed in our study.

Recall that in general equilibrium theory a commodity is defined not only by its physical properties but also by the date, the location, and the state of its nature that precise conditions of its availability. The classical general equilibrium theory deals with a finite number of commodities, which implies that the economic activity extends over only finitely many dates, location, and events. Such an assumption, contrary to the framework of many applied models particularly involving location, growth, and finance, dramatically limits the scope of applications of the results obtained to understand the real economic life.

Moreover, the physical property of a commodity can vary continuously depending on some characteristics. In such dynamic settings, a commodity is defined as a point in a space of characteristics describing, e.g., its location, design, quality, functioning, etc. To give a solid foundation to analysis of allocation over time or state of nature was the primary motivation for the study of infinite-dimensional economies; see, e.g., the paper by Debreu and Hildenbrand [11], which was one of the first to seriously address this and related issues.

Our paper concerns the study of infinite-dimensional nonconvex economic models from the viewpoint of Pareto optimality/efficiency. It has been fully recognized that the concept of Pareto optimality and its variants play a crucial role in equilibrium models to make the best decisions for competitive economies; see, e.g., [2, 3, 9, 10, 17, 20, 31] with detailed discussions and the references therein.

The classical (foundation) approach to the study of Pareto optimality in models of welfare economics with only private goods and smooth data in finite-dimensional spaces of commodities consists of reducing them to conventional problem of mathematical programming and and the subsequent usage of first-order necessary optimality conditions involving Lagrange multipliers; see Hicks [15], Lange [22], and Samuelson [33, 34]. It was shown in this way that the *marginal rates* of substitution in consumption and production sectors are *identical* at any Pareto optimal allocation of resources.

The underlying hypothesis in the foundation works on welfare economics [15, 22, 33, 34] was the *smoothness/differentiability* of the production and utility functions involved in the models. In the beginning of 1950s, Arrow

[3] and Debreu [10] made the next crucial step in general equilibrium theory considering models of welfare economics with possibly nonsmooth but *convex* data in finite-dimensional commodity spaces. Based on the classical separation theorem for convex sets, they established a key result called the *second fundamental theorem of welfare economics*. This theorem, which is a convex counterpart of the aforementioned result on the marginal rates of substitution, states that any Pareto optimal allocation can be *decentralized as price equilibrium*, i.e., it can be sustained by a nonzero price at which each consumer minimizes his/her expenditure and each firm maximizes its profit.

The Arrow–Debreu decentralization approach to Pareto optimality has played a profound role in general equilibrium theory and particularly in welfare economics. On the other hand, the relevance of convexity hypotheses is often doubtful in many applications, e.g., for practical models involving the increasing return to scale in the production sector. In fact, it was observed by Samuelson [34, pp. 231–232], even before the appearance the Arrow–Debreu model, that the convexity assumptions are fulfilled "only be accident."

In his pioneering study on price decentralization of Pareto optimal allocations in nonconvex economies with private goods, Guesnerie [14] imposed convexity assumptions not on the initial production and preference sets, but on their tangential approximations via the Dubovitskii–Milyutin cone of interior displacements. Starting with [14], the results of the second welfare theorem type for nonconvex models are interpreted as *marginal price equilibria*, where marginal prices at Pareto optimal allocations are formalized via the corresponding normal cone that is dual/polar to the tangent cone in question. Guesnerie's approach to nonconvex welfare economics and its elaborations are strongly based on the convexity assumption imposed on the approximating tangent cone.

Further progress in this direction has been achieved by using Clarke's tangent cone in the Guesnerie scheme for various welfare models; see e.g., [6, 9, 21] and their references. An advantage of Clarke's tangent cone is that it is automatically convex, but a strong disadvantage comes from the fact that the corresponding normal cone happens to be too large (often the whole space) and thus does not allow us to bring useful information for marginal price equilibria. These issues have been recognized and discussed by Khan [20].

Khan's new approach to marginal price equilibria goes back, in a sense, to the foundation works in welfare economics [15, 22, 34], which do not use convex separation techniques but reduce the welfare model to an optimization problem that contains at that time nondifferentiable data. He employs, in the case of finite commodities, the metric approximation method developed by Mordukhovich in nondifferentiable programming [26, 27] and arrived in this way, under appropriate constraint qualifications, at the marginal price equilibrium formalized via the Mordukhovich normal cone, which is much smaller than the Clarke one. Further results in this directions for various models of welfare economics can be found in more recent publications [4, 7, 12, 16–18, 23, 28, 29, 31, 37] and their references.

Concerning economies with public goods, the first fundamental result in the "foundation" direction was obtained by Samuelson [35] who showed that at Pareto optimal allocation the marginal rates of transformation of public goods are equal to the sum of the individual marginal rates of substitution. After more than decade from Samuelson's result, Foley [13] and Milleron [25] established appropriate versions of Arrow–Debreu second welfare theorem for economies with public goods under convexity assumptions. More recent results for nonconvex models of welfare economics involving public goods were obtained in [19–21, 36, 37]; see also the references therein.

The main setting of this paper is a general model of nonconvex welfare economics with public goods formulated in the framework of Asplund commodity spaces; see Sect. 2 for mode details. We develop an approach to the study of this model based on advanced tools of variational analysis and generalized differentiation revolving around the *extremal principle*; see [30, Chap. 2]. The latter fundamental principle (in its both approximate and exact forms) provides, on one hand, necessary conditions for a certain extremal relationship between closed sets while, on the other hand, can be treated as a variational nonconvex counterpart of the classical separation principle in convex analysis. This approach was suggested in [28] for welfare models with only private goods and then developed in [4, 5, 23, 29, 31] and other publications. We also refer the reader to [7, 12, 16–18] and the bibliographies therein for similar nonconvex separation ideas closely related to the extremal principle. To some extent, this variational approach via the extremal principle can be viewed as a common roof for the foundation ideas of using first-order optimality conditions as well as for the Arrow–Debreu developments based on convex separation. To the best of our knowledge, such an approach has never been implemented in models of welfare economics with public goods, even in the case of finite commodities.

Developing the variational lines of research, we derive in this paper several new versions of the extended second welfare theorem for nonconvex economies. We start with an approximate/fuzzy version of the second welfare theorem for Pareto and weak Pareto optimal allocations, where marginal prices are formalized via Fréchet/regular normals. Results of this type hold under rather unrestrictive assumptions, but they apply merely to some suboptimal feasible allocations nearby the optimal ones. Imposing additional "normal compactness" requirements on the sets involved in the model, we arrive at an exact/pointwise version of the extended second-welfare theorem with marginal prices formalized via the Mordukhovich normal cone. Furthermore, the usage of the advanced tools of generalized differential and of the associated variational techniques allows us to establish certain *decentralized* versions of the marginal price results by employing some *nonlinear prices*. Considering finally, strong Pareto optimal allocations, the notion introduced by Khan [19] and largely underinvestigated in the literature, we obtain the corresponding versions of the extended second welfare theorem for nonconvex economies with public goods in the absence of the conventional qualification conditions needed for Pareto and weak Pareto optimal allocations.

The rest of the paper is organized as follows. In Sect. 2 we describe and discuss a nonconvex *model* of welfare economics with both private and public goods in infinite-dimensional commodity spaces. We also formulate general *qualification conditions* employed in the paper for studying Pareto and weak Pareto optimal allocations and present some sufficient conditions for their fulfillment.

Section 3 contains required *preliminaries* from variational analysis and generalized differentiation needed for deriving the main results of the paper on the extended versions of the second welfare theorem, which are given in the subsequent sections.

In Sect. 4 we first establish an *approximate* version of the second welfare theorem with *marginal prices* formalized via the Fréchet normal cone at Pareto and weak Pareto optimal allocations under the corresponding *net demand qualification* conditions for nonconvex economies involving public goods. After discussing some remarkable consequences of this result, we derive its *decentralized* (Arrow–Debreu type) version by using nonlinear prices. This is based on a smooth variational description of Fréchet normals.

Section 5 is devoted to *exact* versions of the second welfare theorem for *weak Pareto* and *Pareto* optimal allocations under the so-called *sequential normal compactness* (SNC) conditions imposed on (some of) the sets in question. Conditions of this type are automatic in finite dimensions and always hold for sets with certain Lipschitzian properties in infinite-dimensional spaces and can be viewed as far-going extensions of the classical nonempty interior property of convex sets; we do not require the latter even in convex settings.

The final Sect. 6 concerns welfare economies with *ordered* commodity spaces. First we establish the *price positivity* for such models under natural assumptions of the *desirability/free disposal* type and then derive new results on the fulfillment of an extended version of the second welfare theorem specific for *strong Pareto* optimal allocations of convex and nonconvex economies with private and public goods.

Our notation is basically standard in variational analysis and economic modeling; cf. [20, 30, 31, 37]. Unless otherwise stated, the generic space X under consideration is Banach with the norm $\|\cdot\|$ and the canonical pairing $\langle\cdot,\cdot\rangle$ between X and its topological dual X^* endowed with the weak*

topology w^*. Recall that $I\!B$ and $I\!B^*$ stand for the closed unit ball in X and X^*, respectively; $I\!N := \{1, 2, \ldots\}$ is the collection of natural numbers. Given a set-valued mapping $F\colon X \rightrightarrows X^*$, denote by

$$\operatorname*{Lim\,sup}_{x\to\bar{x}} F(x) := \Big\{x^* \in X^* \Big| \exists \text{ sequences } x_k \to \bar{x},\ x_k^* \xrightarrow{w^*} x^* \text{ as } k \to \infty \text{ with } x_k^* \in F(x_k) \text{ for all } k \in I\!N\Big\} \tag{1.1}$$

the *sequential Painlevé–Kuratowski outer/upper limit* of F as $x \to \bar{x}$. Some more specific symbols are defined in the text below.

2. Nonconvex economies with private and public goods

Let us first formulate a general and well-recognized by now model of welfare economics with private and public goods; see, e.g., [20, 35]. This model denoted by $\mathcal{E}$ involves two categories of commodities: private and public. Consumption of the first type is exclusive, i.e., what is taken by one individual automatically becomes unavailable for all the others. In contrast, a good is public if its consumption is identical across all the individuals. Mathematically this means that the commodity space E is represented as the product of two spaces $E = E_\pi \times E_g$, where E_π stands for the commodity space pertaining to private goods while E_g is the commodity space of public goods.

The consumption set for the ith consumer is given by a subset C^i of E that describes those consumption bundles that can be realized. A consumption plan for the ith consumer is a bundle $x \in C^i$. A consumption plan specifies some amount of goods and labor, which the ith consumer is able to realize; thus the ith consumption set $C^i \subset E$ is the collection of all these consumption plans.

Production refers to a process by which certain commodities (inputs) are transformed into different ones (outputs). A production plan for the jth firm, denoted by $y^j \in E$, specifies the amount of inputs that are required to obtain some outputs. A production set for the jth firm, denoted by $S^j \subset E$, is the collection of all the production plans that are possible for the jth firm according to the technological knowledge available to it.

2.1. The model

The economy $\mathcal{E}$ under consideration consists of $n \in I\!N$ consumers with the corresponding consumption sets $C^i = C^i_\pi \times C^i_g$ as $i = 1, \ldots, n$, where $C^i_\pi \subset E_\pi$ and $C^i_g \subset E_g$, and $m \in I\!N$ production sets $S^j \subset E = E_\pi \times E_g$ as

$j = 1, \ldots, m$. We suppose for simplicity that $C_g^i = C_g$ for all $i = 1, \ldots, n$, i.e., everyone in the consumer sector chooses the same consumer bundles x_g of public goods and different bundles of private goods. The results of this paper show, in particular, that each consumer faces different prices for public goods but the same price for all the private goods of the economy.

For each consumer $i \in \{1, \ldots, n\}$ we have the *preference set*

$$P^i(x) = P_\pi^i(x_\pi) \times P_g^i(x_g) \subset C_\pi^i \times C_g^i$$

defined as the collection of all elements in C^i preferred to x^i by this consumer at the consumption plan $x = (x^i \mid i = 1, \ldots, n)$. It follows from the definition that $x^i \notin P^i(x)$ for all $i = 1, \ldots, n$. In an economy with public goods it is natural to assume that at least one individual desires the public good, i.e., $P_g^i(x_g) \neq \emptyset$ for the corresponding index i. For convenience we put $\operatorname{cl} P^i(x) = \{x^i\}$ if $P^i(x) = \emptyset$. Observe that we do not use utility functions to describe preference sets and also do not impose any preordering and/or other conventional assumptions of welfare economics.

Let us finally consider a general *net demand constraint set* $W = W_\pi \times W_g \subset E$ allowing us to describe market constraints of the economy and unify some conventional situations in economic modeling; cf. [28, 31] for welfare economies with only private goods. In the classical "markets clear" setting we have $W = (\omega, 0)$, where $\omega \in W_\pi$ is an aggregate initial endowment of scarce resources and $0 \in W_g$ means that there is no endowment of public goods. Another conventional situation corresponding to the so-called "implicit free disposal" of private commodities is modeled by $W_g = \{0\}$ and $W_\pi := \omega - (E_\pi)_+$, where $(E_\pi)_+$ stands for the closed positive cone of an ordered private commodity space E_π; see, e.g., [9, 20]. In general the net demand set W describes natural situations that may happen when the initial aggregate endowment is not exactly known due to, e.g., incomplete information. The latter particularly reflects uncertainties in economic modeling.

Next we define the notions of feasible allocations for the economy $\mathcal{E}$ under consideration.

Definition 2.1 (feasible allocations). *We say that* $(x, y) \in \prod_{i=1}^n C^i \times \prod_{j=1}^m S^j$ *is a* FEASIBLE ALLOCATION *of* $\mathcal{E}$ *if the market constraint*

$$\sum_{i=1}^{n} (x_\pi^i, x_g) - \sum_{j=1}^{m} y^j \in W \tag{2.1}$$

is satisfied with the given net demand constraint set W.

The main goal of this paper is to study the following Pareto-type notions of optimal allocations for $\mathcal{E}$ from the viewpoint of deriving necessary optimality conditions for them, which provide in fact extended versions of the second welfare theorem for economies with (private and) public goods.

Definition 2.2 (Pareto-type optimal allocations). *Let* $(\bar{x}, \bar{y})$ *be a feasible allocation of the economy* $\mathcal{E}$ *with the local satiation property*

$$\bar{x}^i \in \operatorname{cl} P^i(\bar{x}) \ \textit{for all} \ i = 1, \ldots, n. \tag{2.2}$$

Then we say that:

(i) $(\bar{x}, \bar{y})$ *is a* WEAK PARETO OPTIMAL ALLOCATION *of* $\mathcal{E}$ *if for any feasible allocation* (x, y) *of* $\mathcal{E}$ *we have the relationships:*

$$x^i \notin P^i(\bar{x}) \ \textit{for some index} \ i \in \{1, \ldots, n\}. \tag{2.3}$$

(ii) $(\bar{x}, \bar{y})$ *is a* PARETO OPTIMAL ALLOCATION *of* $\mathcal{E}$ *if for any feasible allocation* (x, y) *of* $\mathcal{E}$ *we have the relationships:*

$$\textit{either} \ x^i \notin \operatorname{cl} P^i(\bar{x}) \ \textit{for some} \ i \in \{1, \ldots, n\} \ \textit{or} \ x^i \notin P^i(\bar{x}) \\ \textit{for all} \ i = 1, \ldots, n. \tag{2.4}$$

(iii) $(\bar{x}, \bar{y})$ *is a* STRONG PARETO OPTIMAL ALLOCATION *of* $\mathcal{E}$ *if for any feasible allocation* (x, y) *with* $x \neq \bar{x}$ *we have*

$$x^i \notin \operatorname{cl} P^i(\bar{x}) \ \textit{for some} \ i \in \{1, \ldots, n\}. \tag{2.5}$$

It is clear that (iii)$\Longrightarrow$(ii)$\Longrightarrow$(i) but not vice versa. Note the notions of Pareto and weak Pareto optimal allocations are conventional in welfare economics. The notion of strong Pareto optimal allocations was introduced by Khan [19] and then was studied, e.g., in [4, 5, 20, 23, 29, 31], particularly for economies with only private goods.

We can naturally define appropriate *local* versions of the above Pareto-type optimal allocations, but in this paper we confine ourselves with the study of the (global) versions from Definition 2.2. In the recent paper [4] the reader can find more discussions on the relationships between local and global Pareto-type notions and the corresponding versions of the extended second welfare theorem for economies with only private goods.

2.2. Qualification conditions

We formulate and discuss here the mild qualification conditions introduced in [28] and then used in [23, 29, 31] for studying Pareto and weak Pareto optimal allocations of welfare economies with only private goods.

Definition 2.3 (net demand qualification conditions). *Let* $(\bar{x}, \bar{y})$ *be a feasible allocation of the economy* $\mathcal{E}$, *and let*

$$\bar{w} := \sum_{i=1}^{n} (\bar{x}_{\pi}^{i}, \bar{x}_{g}) - \sum_{j=1}^{m} \bar{y}^{j}. \tag{2.6}$$

Given $\varepsilon > 0$, *consider the set*

$$\Delta_{\varepsilon} := \sum_{i=1}^{n} \left[\mathrm{cl}\, P^{i}(\bar{x}) \cap (\bar{x}^{i} + \varepsilon \mathbb{B})\right] - \sum_{j=1}^{m} \left[\mathrm{cl}\, S^{j} \cap (\bar{y}^{j} + \varepsilon \mathbb{B})\right] - \mathrm{cl}\, W \cap (\bar{w} + \varepsilon \mathbb{B}) \tag{2.7}$$

and say that:

(i) *The* NET DEMAND WEAK QUALIFICATION (NDWQ) CONDITION *holds at* $(\bar{x}, \bar{y})$ *if there are a number* $\varepsilon > 0$ *and a sequence* $\{e^k\} \subset E$ *with* $e^k \to 0$ *as* $k \to \infty$ *such that*

$$\Delta_{\varepsilon} + e^{k} \subset \sum_{i=1}^{n} P^{i}(\bar{x}) - \sum_{j=1}^{m} S^{j} - W \quad \textit{for all large } k \in \mathbb{N}. \tag{2.8}$$

(ii) *The* NET DEMAND QUALIFICATION (NDQ) CONDITION *holds at* $(\bar{x}, \bar{y})$ *if there are a number* $\varepsilon > 0$, *a sequence* $\{e^k\} \subset E$ *with* $e^k \to 0$ *as* $k \to \infty$, *and a consumer index* $i_0 \in \{1, \ldots, n\}$ *such that*

$$\Delta_{\varepsilon} + e^{k} \subset P^{i_0}(\bar{x}) + \sum_{i \neq i_0} \mathrm{cl}\, P^{i}(\bar{x}) - \sum_{j=1}^{m} S^{j} - W \quad \textit{for all large } k \in \mathbb{N}. \tag{2.9}$$

Since we obviously have the inclusion

$$\sum_{i=1}^{n} P^{i}(\bar{x}) \subset P^{i_0}(\bar{x}) + \sum_{i \neq i_0} \mathrm{cl}\, P^{i}(\bar{x}),$$

the NDWQ condition implies the NDQ one while, as it is easy to see, not vice versa. Note that for economies with only private goods and markets clear constraints – i.e., when $W = (\omega, 0) \in E_{\pi} \times E_{g}$ in (2.1) – the NDQ condition was defined and applied in [17, 18] with discussing sufficient conditions for its validity that cover those from [6, 9] and other publications. Some extension of the NDQ condition has been recently introduced in [16] for private good economies; see also [4, 5] for further extensions and more discussions. In [13, 19–21, 25, 35, 36] the reader can find qualification conditions for economies with public goods implying the NDQ condition from Definition 2.3.

Observe that the NDQ condition is designed to handle Pareto optimal allocations of welfare economies while the NDWQ one is more appropriate for the study of weak Pareto optimality; see below. We refer the reader to [4, 5, 12] for some specifications of the NDWQ condition for weak Pareto optimal allocations and their relationships with Mas-Collel's uniform properness [24] and its modifications for economies with private goods.

To present next verifiable conditions ensuring the NDQ and NDWQ ones, we recall an important property of sets broadly used in the paper. A nonempty subset $\Omega \subset X$ of a normed space is *epi-Lipschitzian* around $\bar{x} \in \mathrm{cl}\,\Omega$ (in the sense of Rockafellar [32]) if there are neighborhoods U of $\bar{x}$ and O of $0 \in X$, a number $\gamma > 0$, and a vector $c \in X$ such that

$$\Omega \cap U + tO \subset \Omega + tc \quad \text{for all} \quad t \in (0, \gamma). \tag{2.10}$$

If the set Ω is closed around $\bar{x}$, property (2.10) with $c \neq 0$ is equivalent to a local homeomorphic representation of Ω via the epigraph of a real-valued Lipschitz continuous functions; that as where the name comes from. It is worth mentioning that the closure of Ω is epi-Lipschitzian around $\bar{x} \in \mathrm{cl}\,\Omega$ if the set Ω enjoys this property around this point, but not vice versa. Furthermore, a convex set Ω is epi-Lipschitzian around each of its point if and only if $\mathrm{int}\,\Omega \neq \emptyset$; see, e.g., [30, Proposition 1.25].

The following proposition gives sufficient conditions for the fulfillment of the NDWQ and NDQ properties of Definition 2.3.

Proposition 2.4 (sufficient conditions for the validity of net demand constraint qualifications). *Let $(\bar{x}, \bar{y})$ be a feasible allocation of the economy $\mathcal{E}$ with public goods defined above. The following assertions holds:*

(i) *Assume that the sets S^j, $j = 1, \ldots, m$, and W are closed around the points $\bar{y}^j$ and $\bar{w}$ from (2.6), respectively. Then the NDQ condition is satisfied at $(\bar{x}, \bar{y})$ if there are $\varepsilon > 0$, $i \in \{1, \ldots, n\}$, and a desirability sequence $\{e^{ik}\} \subset E$ with $e^{ik} \to 0$ as $k \to \infty$ such that*

$$\mathrm{cl}\, P^i(\bar{x}) \cap (\bar{x}^i + \varepsilon I\!B) + e^{ik} \subset P^i(\bar{x}) \quad \textit{for all large} \quad k \in I\!N. \tag{2.11}$$

Moreover, the NDWQ condition is satisfied at $(\bar{x}, \bar{y})$ if a desirability sequence $\{e^{ik}\}$ in (2.11) exists for each $i \in \{1, \ldots, n\}$ with some $\varepsilon > 0$.

(ii) *Assume that $\bar{x}^i \in \mathrm{cl}\, P^i(\bar{x})$ for all $i \in \{1, \ldots, n\}$. Then the NDWQ condition is satisfied at $(\bar{x}, \bar{y})$ if the set*

$$\Delta := \sum_{i=1}^{n} P^i(\bar{x}) - \sum_{j=1}^{m} S^j - W \tag{2.12}$$

is epi-Lipschitzian at $0 \in \mathrm{cl}\,\Delta$. It happens, in particular, when either one among the sets $P^i(\bar{x})$ for $i = 1, \ldots, n$, S^j for $j = 1, \ldots, m$, and W or *some of their partial combinations in* (2.12) *is epi-Lipschitzian around the corresponding point.*

(iii) *Assume that* $n > 1$. *The NDQ condition is satisfied at* $(\bar{x}, \bar{y})$ *if there is a consumer* $i_0 \in \{1, \ldots, n\}$ *such that* $P^{i_0}(\bar{x}) \neq \emptyset$ *and the set*

$$\Sigma := \sum_{i \neq i_0} \mathrm{cl}\, P^i(\bar{x}) \tag{2.13}$$

is epi-Lipschitzian at the point $\sum_{i \neq i_0} \bar{x}^i$. *It happens, in particular, when either one among the sets* $P^i(\bar{x})$ *for* $i \in \{1, \ldots, n\} \setminus \{i_0\}$ *or some of their partial combinations in* (2.13) *is epi-Lipschitzian around the corresponding point.*

Proof. Similar to that in [31, Proposition 8.4] for economies with private goods. △

Note that condition (2.11) is a direct generalization of the desirability direction condition in [24], which is related to the classical "more is better" assumption for convex economies with only private goods and commodity spaces ordered by their closed positive cones having nonempty interiors. Furthermore, it is important to observe that we do not need to impose any assumption on the preference and/or production sets for the validity of both qualification conditions in Definition 2.3 if the net demand constraint set W is epi-Lipschitzian around $\bar{w}$; this easily follows from Proposition 2.4(ii). The latter covers, in particular, the case of free-disposal Pareto optimum; see, e.g., [9, 20].

3. Tools of variational analysis

This section contains some constructions and preliminary results from variational analysis and generalized differentiation that are widely used in this paper to derive extended versions of the second welfare theorem in nonconvex economies with public goods. We mostly follow the book [30], where the reader can find all the proofs and more discussions.

3.1. Generalized normals

We start with constructions of generalized normals to subsets of Banach spaces.

Definition 3.1 (generalized normals to sets). *Let* Ω *be a nonempty subset of* E.

(i) *Given* $x \in \Omega$ *and* $\varepsilon \geq 0$*, the* SET OF ε-NORMALS *to* Ω *at* $x \in \Omega$ *is defined by*

$$\widehat{N}_\varepsilon(x;\Omega) := \Big\{x^* \in E^* \Big|\ \operatorname*{Lim\,sup}_{u \xrightarrow{\Omega} x} \frac{\langle x^*, u - x\rangle}{\|u - x\|} \leq \varepsilon\Big\}, \tag{3.1}$$

where Limsup *stands for the Painlevé–Kuratowski outer limit* (1.1) *and where the symbol* $u \xrightarrow{\Omega} x$ *means that* $u \to x$ *with* $u \in \Omega$*. When* $\varepsilon = 0$ *in* (3.1)*, the cone* $\widehat{N}(x;\Omega) := \widehat{N}_0(x;\Omega)$ *is called the* FRÉCHET NORMAL CONE *to* Ω *at* x.

(ii) *Given* $\bar{x} \in \Omega$*, the outer limit*

$$N(\bar{x};\Omega) := \operatorname*{Lim\,sup}_{\substack{x \xrightarrow{\Omega} \bar{x} \\ \varepsilon \downarrow 0}} \widehat{N}_\varepsilon(x;\Omega) \tag{3.2}$$

is called the M(ORDUKHOVICH)-NORMAL CONE *to* Ω *at* $\bar{x}$.

Note that construction (3.1) with $\varepsilon = 0$ is also known in the literature as the *prenormal* or *regular normal* cone while (3.2) as *basic* or *limiting normal cone*. If Ω is locally closed around $\bar{x}$ and the space E is *Asplund* (i.e., each of its separable subspace has a separable dual), then the normal cone (3.2) admits the simplified representation

$$N(\bar{x};\Omega) := \operatorname*{Lim\,sup}_{x \xrightarrow{\Omega} \bar{x}} \widehat{N}(x;\Omega). \tag{3.3}$$

The class of Asplund spaces is sufficiently large including, in particular, every reflexive Banach space and every space with a separable dual; see [30] for more details and references. If $E = I\!R^n$ and Ω is locally closed around $\bar{x}$, representation (3.3) is equivalent to the original definition in [26] given by

$$N(\bar{x};\Omega) = \operatorname*{Lim\,sup}_{x \to \bar{x}} \big[\operatorname{cone}\big(x - \Pi(x;\Omega)\big)\big] \tag{3.4}$$

via the Euclidean projector $\Pi(x;\Omega)$ of x on Ω, where the symbol 'cone' signifies the conic hull spanned on the set in question.

In the case of convex sets we have the following representations of generalized normals from Definition 3.1 showing, in particular, that both cones $\widehat{N}(\bar{x};\Omega)$ and $N(\bar{x};\Omega)$ under consideration extend the classical one in convex analysis.

Proposition 3.2 (normals to convex sets). *Let* Ω *be convex. Then given arbitrary* $\bar{x} \in \Omega$ *and* $\varepsilon \geq 0$*, we have*

$$\widehat{N}_\varepsilon(\bar{x};\Omega) = \widehat{N}(\bar{x};\Omega) + \varepsilon I\!B^* = \big\{x^* \in E^* \big|\ \langle x^*, x - \bar{x}\rangle \leq \varepsilon\|x - \bar{x}\| \ \text{for all}\ x \in \Omega\big\}, \tag{3.5}$$

which implies the representations

$$\widehat{N}(\bar{x};\Omega) = N(\bar{x};\Omega) = \left\{x^* \in E^* \,\middle|\, \langle x^*, x - \bar{x}\rangle \le 0 \text{ for all } x \in \Omega\right\} \quad (3.6)$$

Note that in the general nonconvex case the first equality of (3.5) is replaced by

$$\widehat{N}_\varepsilon(\bar{x};\Omega) \supset \widehat{N}(\bar{x};\Omega) + \varepsilon I\!B^*, \quad \varepsilon > 0.$$

In what follows we also employ the following useful formulas for representing generalized normals to products of arbitrary sets.

Proposition 3.3 (normal cones to products of sets). *Let $\Omega_i \subset E_i$, $i = 1, 2$, be nonempty subsets of Banach spaces, and let $\bar{x} = (\bar{x}_1, \bar{x}_2) \in \Omega_1 \times \Omega_2$. Then we have*

$$\widehat{N}(\bar{x};\Omega_1 \times \Omega_2) = \widehat{N}(\bar{x}_1;\Omega_1) \times \widehat{N}(\bar{x}_2;\Omega_2),$$
$$N(\bar{x};\Omega_1 \times \Omega_2) = N(\bar{x}_1;\Omega_1) \times N(\bar{x}_2;\Omega_2).$$

The next important results taken from [30, Theorem 1.30] provide *smooth variational* representations of Fréchet normals that play a crucial role in nonlinear price descriptions of *decentralized equilibria* in the extended second welfare theorems for economies with public good established in this paper. Note that assertion (i) of Theorem 3.4 below holds in arbitrary Banach space, while assertion (ii) gives an essentially stronger description of Fréchet normals under an additional geometric assumption on the space in question.

Recall that a Banach space E is *Fréchet smooth* if there is an equivalent norm on E that is Fréchet differentiable at any nonzero point. In particular, every reflexive space is Fréchet smooth. Observe also that every Fréchet smooth space is Asplund, and hence we can use formula (3.3) for representing our basic normals.

Theorem 3.4 (smooth variational descriptions of Fréchet normals). *Let Ω be a nonempty subset of a Banach space E, and let $\bar{x} \in \Omega$. The following assertions hold:*

- **(i)** *Given $x^* \in E^*$, assume that there is a function $s\colon U \to I\!R$ defined on a neighborhood of $\bar{x}$ and Fréchet differentiable at $\bar{x}$ such that $\nabla s(\bar{x}) = x^*$ and $s(x)$ achieves a local maximum relative to Ω at $\bar{x}$. Then $x^* \in \widehat{N}(\bar{x};\Omega)$. Conversely, for every $x^* \in \widehat{N}(\bar{x};\Omega)$ there is a function $s\colon E \to I\!R$ such that $s(x) \le s(\bar{x}) = 0$ whenever $x \in \Omega$ and that $s(\cdot)$ is Fréchet differentiable at $\bar{x}$ with $\nabla s(\bar{x}) = x^*$.*
- **(ii)** *Assume that E is Fréchet smooth. Then for every $x^* \in \widehat{N}(\bar{x};\Omega)$ there is a concave Fréchet differentiable function $s\colon E \to I\!R$ that achieves its global maximum relative to Ω uniquely at $\bar{x}$ and such that $\nabla s(\bar{x}) = x^*$.*

One of the major features of the M-normal cone (3.2) is its *nonconvexity*, even in the case of rather simple nonconvex sets in finite dimensions, e.g., when Ω is either the graph of the function $|x|$ at $(0, 0) \in I\!R^2$ or the epigraph of the function $-|x|$ at the origin. This does not allows us to employ conventional techniques of convex and 'convexified' analysis, mainly based on convex separation theorems and related results, to the study and applications of the M-normal cone and the associated constructions for functions and (single-valued and set-valued) mappings. Also the nonconvexity of (3.2) indicates that this normal cone is not dual/polar to any tangent cone, since polarity always implies convexity. Nevertheless, the M-normal cone and the corresponding subdifferential and coderivative constructions enjoy *full calculus* and other nice properties crucial in applications, mainly in the general framework of Asplund spaces; see [30, 31] and the references therein. These phenomena are based on advanced *variational/extremal principles* of modern variational analysis.

In the next subsection we present the basic extremal principle in Asplund space used in this paper for deriving extended versions of the second welfare theorem for nonconvex economies with infinite commodities and public goods.

3.2. Extremal principle

First we present and briefly discuss the required definitions and then formulate the underlying results on the extremal principle. The reader can find full proofs, more discussions, and references in [30, Chap. 2] and the commentaries therein.

Definition 3.5 (local extremal points). *Let $\Omega_1, \ldots, \Omega_n$ with $n \geq 2$ be nonempty subsets of a normed space E. We say that $\bar{x} \in \cap_{i=1}^n \Omega_i$ is a* LOCAL EXTREMAL POINT *of the set system $\{\Omega_1, \ldots, \Omega_n\}$ if there are sequences $\{a_i^k\} \subset E$, $i = 1, \ldots, n$, and a neighborhood U of $\bar{x}$ such that $a_i^k \to 0$ as $k \to \infty$ and that*

$$\bigcap_{i=1}^{n}(\Omega_i - a_i^k) \cap U = \emptyset \quad \text{for all large } k \in I\!N.$$

The system of sets having at least one local extremal point is called an EXTREMAL SYSTEM.

As shown in [30, 31] and their references, the concept of set extremality encompasses various notions of optimal solutions to problems of scalar and vector/multiobjective optimization. On the other hand, a number of nonvariational issues (e.g., calculus rules, stability, etc.) reduce to extremal systems of sets by using a variational approach. In this paper we show that the above notions of Pareto optimal allocations in economies with public goods can be reduced to extremal points of appropriate set systems under the imposed net demand qualification conditions for the cases Pareto and weak Pareto optimal allocations and with no such conditions for strong Pareto ones.

The next theorem contains two versions of the extremal principle used in what follows. The first version is approximate, which does not require any extra assumptions but expresses the result in terms of Fréchet normals to the sets Ω_i at points nearby the local extremal one. The second version provides an exact extremality condition formulated via the M-normal cone at the local extremal point in question under certain additional assumptions on the sets Ω_i. These assumptions are automatic in finite dimensions while imposing a sufficient amount of "normal compactness" in infinite-dimensional spaces. In the sequel we use the following condition, perhaps the weakest one of this type needed for general systems of sets.

Definition 3.6 (sequential normal compactness). *Let $\Omega \subset E_1 \times E_2$ be a set in the product of two normed spaces. We say that:*

(i) Ω *is* SEQUENTIALLY NORMALLY COMPACT (SNC) *at* $\bar{x} \in \Omega$ *if for any sequences* $\varepsilon_k \downarrow 0$, $x_k \to \bar{x}$ *with* $x_k \in \Omega$ *satisfying*

$$(x_k^{*1}, x_k^{*2}) \in \widehat{N}_{\varepsilon_k}(x_k; \Omega) \ \text{for all} \ k \in I\!N$$

we have the implication

$$(x_k^{*1}, x_k^{*2}) \overset{w^*}{\to} (0,0) \Longrightarrow \|(x_k^{*1}, x_k^{*2})\| \to 0 \ \text{as} \ k \to \infty.$$

(ii) Ω *is* PARTIALLY SNC (PSNC) *at* $\bar{x}$ *with respect to the first component if for any sequences* $(\varepsilon_k, x_k, x_k^{*1}, x_k^{*2})$ *from* (i) *we have the implication*

$$\big[x_k^{*1} \overset{w^*}{\to} 0,\ \|x_k^{*2}\| \to 0\big] \Longrightarrow \|x_k^{*1}\| \to 0 \ \text{as} \ k \to \infty.$$

(iii) Ω *is* STRONGLY PSNC *at* $\bar{x}$ *with respect to the first component if for any sequences* $(\varepsilon_k, x_k, x_k^{*1}, x_k^{*2})$ *from* (i) *we have the implication*

$$\big[(x_k^{*1}, x_k^{*2}) \overset{w^*}{\to} (0,0)\big] \Longrightarrow \|x_k^{*1}\| \to 0 \ \text{as} \ k \to \infty.$$

If Ω is a locally closed subset of an Asplund space, we can equivalently put $\varepsilon_k \equiv 0$ in all the relationships of Definition 3.6. Observe that, besides sets in finite-dimensional spaces, the SNC property holds for any subset of a

Banach space that is *compactly epi-Lipschitzian* (CEL) around $\bar{x} \in \Omega$ in the sense of of Borwein and Strójwas [8], which means that a singleton $\{c\}$ in definition (2.10) of the epi-Lipschitzian property is replaced by some compact set $C \subset E$. More subtle conditions of the Lipschitzian type ensuring the PSNC and strong PSNC properties can be found in [30].

Now we are ready to formulate both approximate and exact versions of the extremal principle proved in [30, Theorem 2.20 and Theorem 2.22], respectively.

Theorem 3.7 (extremal principle in Asplund spaces). *Let $\bar{x}$ be a local extremal point of the set system $\{\Omega_1, \ldots, \Omega_n\}$ in an Asplund space E. Assume that all the sets Ω_i are locally closed around $\bar{x}$. Then the following assertions hold:*

(i) *For any $\varepsilon > 0$ there are $x_i \in \Omega_i \cap (\bar{x} + \varepsilon I\!B)$ and $x_i^* \in E^*$ satisfying the relationships*

$$x_i^* \in \widehat{N}(x_i; \Omega_i) + \varepsilon I\!B^* \quad \text{for all} \quad i = 1, \ldots, n, \tag{3.7}$$

$$\sum_{i=1}^{n} x_i^* = 0, \tag{3.8}$$

$$\sum_{i=1}^{n} \|x_i^*\| = 1. \tag{3.9}$$

(ii) *In addition to the assumptions above, suppose that all but one of the sets Ω_i as $i = 1, \ldots, n$ are SNC at $\bar{x}$. Then there are M-normals*

$$x_i^* \in N(\bar{x}; \Omega_i) \quad \text{for all} \quad i = 1, \ldots, n \tag{3.10}$$

satisfying the relationships in (3.8) *and* (3.9).

It is easy to see from Proposition 3.2 that for the case of two convex sets Ω_1 and Ω_2 relationships (3.8)–(3.10) of the exact extremal principle reduce to conventional convex separation theorem, where the SNC requirement imposed on one of the sets is a far-going extension of the classical interiority condition, even in the case of convexity. Relationships (3.7)–(3.9) of the approximate extremal principle can be treated as a nonconvex counterpart of the celebrated Bishop–Phelps density theorem for convex sets with empty interiors; see [30, Sect. 2.1] for more discussions.

Combining these observations with the fact that the extremal principle gives necessary conditions for set extremality and thus can be considered an "extended Lagrange multipliers rule," our extremal principle approach to the second welfare theorem provides a unification of the classical foundation approach and the Arrow–Debreu separation/decomposition approach to welfare economics in general nonconvex and nonsmooth settings.

4. Approximate versions of the extended second welfare theorem for economies with public goods

In this section we derive necessary optimality conditions for Pareto and weak Pareto optimal allocations of welfare economies with public goods given in certain approximate forms under the underlying NDQ and NDWQ qualification conditions. The essence of such results, which provide extended versions of the second welfare theorem for convex and nonconvex economies, is that their formulations involve not only the optimal allocation in question, but also feasible allocations nearby. The results obtained in approximate forms hold with no extra assumptions on the initial data in infinite-dimensional commodity spaces.

We present two versions of the approximate second welfare theorem for economies with public goods mentioned in Sect. 1. Let us start with the first version, where (linear) *marginal prices* are formalized via Fréchet normals. The results are given and proved in a parallel way for Pareto and weak Pareto optimal allocations.

Theorem 4.1 (extended second welfare theorem via approximate marginal prices). *Let $(\bar{x}, \bar{y})$ be a Pareto (resp. weak Pareto) optimal allocation of the economy $\mathcal{E}$ with public goods. Assume that the commodity space E is Asplund and that the NDQ (resp. NDWQ) condition is satisfied at $(\bar{x}, \bar{y})$. Then given any $\varepsilon > 0$, there are a commodity bundle $(x, y, w) \in \prod_{i=1}^{n} \operatorname{cl} P^i(\bar{x}) \times \prod_{j=1}^{m} \operatorname{cl} S^j \times \operatorname{cl} W$ as well as marginal prices $(p_\pi^*, p_g^*) \in E_\pi^* \times E_g^*$ and $p_g^{*i} \in E_g^*$ for $i = 1, \ldots, n$ satisfying the relationships:*

$$\sum_{i=1}^{n} p_g^{i*} = p_g^*, \tag{4.1}$$

$$-(p_\pi^*, p_g^{*i}) \in \widehat{N}\big(x^i; \operatorname{cl} P^i(\bar{x})\big) + \varepsilon \mathbb{B}^*, \quad x^i \in \bar{x}^i + \varepsilon \mathbb{B}, \quad i = 1, \ldots, n, \tag{4.2}$$

$$(p_\pi^*, p_g^*) \in \widehat{N}(y^j; \operatorname{cl} S^j) + \varepsilon \mathbb{B}^*, \quad y^j \in \bar{y}^j + \varepsilon \mathbb{B}, \quad j = 1, \ldots, m, \tag{4.3}$$

$$(p_\pi^*, p_g^*) \in \widehat{N}(w; \operatorname{cl} W) + \varepsilon \mathbb{B}^*, \quad w \in \bar{w} + \varepsilon \mathbb{B}, \tag{4.4}$$

$$1 - \varepsilon/2 \le \max\big\{\|(p_\pi^*, p_g^{*i})\|_{i \in \{1,\ldots,n\}}, \|p_g^*\|\big\} \le 1 + \varepsilon/2 \tag{4.5}$$

Proof. Define the product space $X := E^{n+m+1}$ equipped with the sum norm

$$\big\|(x_1, \ldots, x_{n+m+1})\big\| := \|x_1\| + \ldots + \|x_{n+m+1}\|.$$

In this case the corresponding dual norm on X^* is given by

$$\big\|(x_1^*, \ldots, x_{n+m+1}^*)\big\| = \max\big\{\|x_1^*\|, \ldots, \|x_{n+m+1}^*\|\big\}. \tag{4.6}$$

Observe that the space X is Asplund as a product of Asplund spaces.

To prove both results formulated in the theorem (for Pareto and weak Pareto optimal allocations), we use the *approximate extremal principle* in the Asplund space X applied to the system of the two closed sets defined as follows:

$$\Omega_1 := \prod_{i=1}^{n}\Big[\operatorname{cl} P^i(\bar{x}) \cap (\bar{x}^i + \varepsilon I\!B)\Big] \times \prod_{j=1}^{m}\Big[\operatorname{cl} S^j \cap (\bar{y}^j + \varepsilon I\!B)\Big]$$
$$\times \left[\operatorname{cl} W \cap (\bar{w} + \varepsilon I\!B)\right], \tag{4.7}$$

$$\Omega_2 := \left\{(x, y, w) \in X \Big| \sum_{i=1}^{n}\left(x_\pi^i, x_g\right) - \sum_{j=1}^{m} y^j - w = 0\right\}. \tag{4.8}$$

Let us check that $(\bar{x}, \bar{y}, \bar{w})$ is an *extremal point* of the set system $\{\Omega_1, \Omega_2\}$ defined in (4.7) and (4.8). Indeed, we have by (2.1) and (2.2) that $(\bar{x}, \bar{y}, \bar{w}) \in \Omega_1 \cap \Omega_2$. It remains to show that there is a sequence $\{a^k\} \subset X$ with $a^k \to 0$ as $k \to \infty$ such that

$$(\Omega_1 - a^k) \cap \Omega_2 = \emptyset \tag{4.9}$$

in the case of the Pareto (resp. weak Pareto) optimal allocation $(\bar{x}, \bar{y})$ of $\mathcal{E}$ under the fulfillment of the NDQ (resp. NDWQ) qualification requirements from Definition 2.3.

Let $\varepsilon > 0$ and $\{e^k\} \subset E$ be such that the corresponding condition (2.8) and (2.9) is satisfied. Form a sequence $\{a^k\} \subset X$ by

$$a^k := (0, \ldots, 0, e^k) \ \text{ for all } \ k \in I\!N \tag{4.10}$$

and get by construction that $a^k \to 0$ as $k \to \infty$. Arguing by contradiction, suppose that relationship (4.9) does not hold along a subsequence of $k \to \infty$. Then we find triples $(x^k, y^k, w^k) \in \Omega_1$ such that $(x^k, y^k, w^k) - a^k \in \Omega_2$ for the corresponding numbers $k \in I\!N$. It follows from the constructions in (4.7), (4.8), and (4.10) that

$$\begin{aligned}
&x^{ki} \in \operatorname{cl} P^i(\bar{x}) \cap (\bar{x}^i + \varepsilon I\!B), \quad i = 1, \ldots, n,\\
&y^{kj} \in \operatorname{cl} S^j \cap (\bar{y}^j + \varepsilon I\!B), \quad j = 1, \ldots, m,\\
&w^k \in \operatorname{cl} W \cap (\bar{w} + \varepsilon I\!B), \quad \text{and}\\
&\sum_{i=1}^{n}\left(x_\pi^{ki}, x_g^k\right) - \sum_{j=1}^{m} y^{kj} - w^k + e^k = 0
\end{aligned} \tag{4.11}$$

for $k \in I\!N$ sufficiently large. Comparing (4.11) with (2.8), we have

$$0 \in \sum_{i=1}^{n} P^i(\bar{x}) - \sum_{j=1}^{m} S^j - W,$$

which contradicts the weak Pareto optimality of the allocation $(\bar{x}, \bar{y})$. The comparison of (4.11) with (2.9) gives us the inclusion

$$0 \in P^{i_0}(\bar{x}) + \sum_{i \neq i_0}^{n} \operatorname{cl} P^i(\bar{x}) - \sum_{j=1}^{m} S^j - W, \tag{4.12}$$

which contradicts the Pareto optimality of $(\bar{x}, \bar{y})$. Thus $(\bar{x}, \bar{y}, \bar{w})$ is an extremal point of the set system $\{\Omega_1, \Omega_2\}$ in both cases under consideration.

Applying the extremal principle from assertion (i) of Theorem 3.7 to the system $\{\Omega_1, \Omega_2\}$ at $(\bar{x}, \bar{y}, \bar{w})$, for any $\varepsilon > 0$ we find elements $u = (x^1, \ldots, x^n, y^1, \ldots, y^m, w) \in \Omega_1$, $v \in \Omega_2$, $x^* \in X^*$ with $\|x^*\| = 1$, and

$$u^* \in \widehat{N}(u; \Omega_1), \quad v^* \in \widehat{N}(v; \Omega_2) \tag{4.13}$$

satisfying the relationships

$$\begin{aligned}
&x^* - u^* \in (\varepsilon/2)I\!B^*, \quad -x^* - v^* \in (\varepsilon/2)I\!B^*,\\
&x^i \in \bar{x}^i + (\varepsilon/2)I\!B \ \text{ for } i = 1, \ldots, n,\\
&y^j \in \bar{y}^j + (\varepsilon/2)I\!B \ \text{ for } j = 1, \ldots, m,\\
&w \in \bar{w} + (\varepsilon/2)I\!B, \quad \text{and}\\
&1 - \varepsilon/2 \le \|v^*\| \le 1 + \varepsilon/2, \quad \|v^* + u^*\| \le \varepsilon.
\end{aligned} \tag{4.14}$$

Since the set Ω_2 in (4.8) is a linear subspace of X, we get by (3.6) and the choice of (v, v^*) in (4.13) the equality

$$\langle v^*, \theta \rangle = \langle v^*, v \rangle \ \text{ for all } \theta \in \Omega_2. \tag{4.15}$$

Let us now show that there are prices $p_\pi^* \in E_\pi^*$ and $p_g^{*i} \in E_g^*$ for $i = 1, \ldots, n$ such that the vector v^* in (4.15) admits the representation

$$\begin{aligned}
v^* = \Big(&(p_\pi^*, p_g^{*1}), \ldots, (p_\pi^*, p_g^{*n}), -(p_\pi^*, p_g^*), \ldots, -(p_\pi^*, p_g^*),\\
&-(p_\pi^*, p_g^*) \Big),
\end{aligned} \tag{4.16}$$

where p_g^* is given by (4.1). Indeed, writing $v^* \in (E^{n+m+1})^*$ as

$$v^* = \big((x_\pi^{*1}, x_g^{*1}), \ldots, (x_\pi^{*n}, x_g^{*n}), (y_\pi^{*1}, y_g^{*1}), \ldots, (y_\pi^{*m}, y_g^{*m}), (w_\pi^*, w_g^*)\big), \tag{4.17}$$

it is not hard to observe from (4.15) and the structure of the set Ω_2 in (4.8) that

$$\sum_{i=1}^{n}\langle x_g^{*i}, z\rangle + \langle w_g^*, z\rangle = 0,$$

$$\langle x_\pi^{*i}, z\rangle + \langle y_\pi^{*j}, z\rangle = 0, \quad \langle x_\pi^{*i}, z\rangle + \langle w_\pi^*, z\rangle = 0$$
$$\text{for all } i = 1,\ldots,n \text{ and } j = 1,\ldots,m$$

whenever vector $z \in E_g$ is chosen. The latter clearly gives that

$$w_g^* = -\sum_{i=1}^{n} x_g^{*i} \text{ and } x_\pi^{*i} = -y_\pi^{*j} = -w_\pi^* \text{ for all such } i, j. \tag{4.18}$$

In the same way we get that

$$\sum_{i=1}^{n}\langle x_g^{*i}, z\rangle + \langle y_g^{*j}, z\rangle = 0 \text{ for all } j \in \{1,\ldots,m\} \text{ and } z \in E_g,$$

which implies the relationships

$$y_g^{*j} = -\sum_{i=1}^{n} x_g^{*i} \text{ as } j = 1,\ldots,m. \tag{4.19}$$

Taking into account that x_π^{*i}, y_π^{*j}, and y_g^{*j} are in fact independent of i and j by (4.18) and (4.19), respectively, and denoting

$$p_\pi^* := x_\pi^{*i}, \quad p_g^* := y_g^{*j}, \text{ and } p_g^{*i} := x_g^{*i} \text{ for } i = 1,\ldots,n,$$

we get from (4.17), (4.18), and (4.19) the claimed representation (4.16) of the normal vector v^* from (4.18), where the public goods price p_g^* satisfies (4.1).

It follows from the first relationship in (4.14), representation (4.16), and the form of the dual norm (4.6) that the nontriviality estimates in (4.5) hold. From the second relationship in (4.14) we further get that

$$-v^* \subset u^* + (\varepsilon/2)I\!B^* \subset \widehat{N}(u;\Omega_1) + \varepsilon I\!B^*,$$

which implies by (4.16) that

$$-v^* = \big(-(p_\pi^*, p_g^{*1}),\ldots,-(p_\pi^*, p_g^{*n}), (p_\pi^*, p_g^*),\ldots,(p_\pi^*, p_g^*), (p_\pi^*, p_g^*)\big)$$
$$\in \widehat{N}(u;\Omega_1) + \varepsilon I\!B^*.$$

Employing the first product formula in Proposition 3.3 to the set Ω_1 in (4.7) and observing that all the above triples (x^i, y^j, w) belong to the corresponding ε-neighborhoods of the optimal one $(\bar{x}, \bar{y}, \bar{w})$, we conclude from the latter inclusion that all the relationships in (4.2)–(4.4) are satisfied, which completes the proof of the theorem. △

Observe that the equality relationship (4.1) between the marginal prices in the public goods sector confirms and extends the fundamental conclusion of welfare economics with public goods that goes back to Samuelson [35]: *the marginal rates of transformation for public goods equal to the sum of the individual marginal rates of substitution* at Pareto and weak Pareto optimal allocations.

Let us present a specification of Theorem 4.1 in the conventional case with no initial endowment of public goods, i.e., when $W = W_\pi \times \{0\}$.

Corollary 4.2 (approximate marginal prices for economies with no initial endowment of public goods). *Let $(\bar{x}, \bar{y})$ be a Pareto (resp. weak Pareto) optimal allocation of the economy $\mathcal{E}$ with $W = W_\pi \times \{0\}$, $W_\pi \subset E_\pi$, under the corresponding assumptions of Theorem* 4.1. *Then for any $\varepsilon > 0$ there are $(x, y, w) \in \prod_{i=1}^n \operatorname{cl} P^i(\bar{x}) \times \prod_{j=1}^m \operatorname{cl} S^j \times \operatorname{cl} W$ as well as marginal prices $(p_\pi^*, p_g^*) \in E_\pi^* \times E_g^*$ and $p_g^{*i} \in E_g^*$ for $i = 1, \ldots, n$ satisfying relationships* (4.1)–(4.3) *and* (4.5) *with the replacement of* (4.4) *by*

$$p_\pi^* \in \widehat{N}(w_\pi; \operatorname{cl} W_\pi) + \varepsilon I\!B^*, \quad w_\pi \in \bar{w}_\pi + \varepsilon I\!B.$$

Proof. Follows directly from Theorem 4.1 and Proposition 3.3. △

The next consequence of Theorem 4.1 gives a specification of the results in the case of *convexity* assumptions imposed on preference and production sets. In this case the marginal price relationships reduce, respectively, to *global minimization* (*maximization*) of the *perturbed* consumer expenditures (firm profits) over the corresponding preference (production) sets. This provides an approximate *decentralized price equilibrium* in convex models with no standard interiority assumptions.

Corollary 4.3 (approximate decentralized equilibrium in convex economies with public goods). *In the framework of Theorem* 4.1, *assume that the preference sets* $\operatorname{cl} P^i(\bar{x})$, $i = 1, \ldots, n$, *and the production sets* S^j, $j = 1, \ldots, m$, *are convex. Then for any $\varepsilon > 0$ there exist $(x, y, w) \in \prod_{i=1}^n \operatorname{cl} P^i(\bar{x}) \times \prod_j^m \operatorname{cl} S^j \times \operatorname{cl} W$ and prices $(p_\pi^*, p_g^*) \in E_\pi^* \times E_g^*$ and $p_g^{*i} \in E_g^*$ for $i = 1, \ldots, n$ satisfying relationships* (4.1), (4.4), (4.5), *and the following ones:*

$$\langle (p_\pi^*, p_g^{*i}), (u, v) - (x_\pi^i, x_g) \rangle \geq -\varepsilon \|(u, v) - (x_\pi^i, x_g)\| \text{ for all } (u, v) \in \operatorname{cl} P^i(\bar{x}),\ i = 1, \ldots, n,$$

$$\langle (p_\pi^*, p_g^*), (u, v) - (y_\pi^j, y_g^j) \rangle \leq \varepsilon \|(u, v) - (y_\pi^j, y_g^j)\| \text{ for all } (u, v) \in \operatorname{cl} S^j,\ j = 1, \ldots, m.$$

Proof. Follows from conditions (4.2) and (4.3) of Theorem 4.1 and the representation of ε-normals to convex sets given in Proposition 3.2. △

The next theorem, developing the corresponding results of [29, 31] to economies with public goods, establishes a *decentralized price equilibrium* of the convex type as in Corollary 4.3 but for general *nonconvex* models. The "price to pay" for this is the usage of *nonlinear prices* in nonconvex models instead of conventional linear prices as in Theorem 4.1 and Corollary 4.3. Note that the essence of nonlinear prices used here as well as in [29, 31] is different from that of [2] and related publications dealing with convex economies.

Theorem 4.4 (decentralized approximate equilibrium in nonconvex models with public goods via nonlinear prices). *Given any $\varepsilon > 0$, the following assertions hold:*

(i) *Let all the assumptions of Theorem 4.1 be fulfilled for a Pareto (resp. weak Pareto) optimal allocation $(\bar{x}, \bar{y})$ of the economy $\mathcal{E}$. Then there exist a commodity bundle $(x, y, w) \in \prod_{i=1}^{n} \mathrm{cl}\, P^i(\bar{x}) \times \prod_{j}^{m} \mathrm{cl}\, S^j \times \mathrm{cl}\, W$, marginal prices $(p_\pi^*, p_g^*) \in E_\pi^* \times E_g^*$ and $p_g^{*i} \in E_g^*$ as $i = 1, \ldots, n$ satisfying relationships (4.1), (4.4), and (4.5) as well as real-valued functions f^i as $i = 1, \ldots, n$ and h^j as $j = 1, \ldots, m+1$ on E that are Fréchet differentiable at x^i, y^j, and w, respectively, with*

$$\begin{cases} \|\nabla f^i(x^i) - (p_\pi^*, p_g^{*i})\| \le \varepsilon, & i = 1, \ldots, n, \\ \|\nabla h^j(y^j) - (p_\pi^*, p_g^*)\| \le \varepsilon, & j = 1, \ldots, m, \\ \|\nabla h^{m+1}(w) - (p_\pi^*, p_g^*)\| \le \varepsilon \end{cases} \tag{4.20}$$

and such that each f^i, $i = 1, \ldots, n$, achieves its global minimum over $\mathrm{cl}\, P^i(\bar{x})$ at x^i, each h^j, $j = 1, \ldots, m$, achieves its global maximum over $\mathrm{cl}\, S^j$ at y^j, and h^{m+1} achieves its global maximum over $\mathrm{cl}\, W$ at w.

(ii) *If in addition to the assumptions in (i) the commodity space E is Fréchet smooth, then the nonlinear prices f^i and h^j can be chosen to be Fréchet differentiable on E and such that each f^i, $i = 1, \ldots, n$, is convex and achieves its global minimum over $\mathrm{cl}\, P^i(\bar{x})$ uniquely at x^i while each h^j, $j = 1, \ldots, m+1$, is concave and achieves its global maximum over $\mathrm{cl}\, S^j$ for $j = 1, \ldots, m$ and over $\mathrm{cl}\, W$ for $j = m+1$ uniquely at y^j and w, respectively.*

Proof. Take the marginal prices $(p_\pi^*, p_g^*, p_g^{i*})$ satisfying all the conclusions of Theorem 4.1 and then, by (4.2) and (4.3), find $(\widetilde{p}_\pi^*, \widetilde{p}_g^*, \widetilde{p}_g^{i*})$ such that

$$\begin{cases} -(\widetilde{p}_\pi^*, \widetilde{p}_g^{*i}) \in \widehat{N}(x^i; \mathrm{cl}\, P^i(\bar{x})), & \|(\widetilde{p}_\pi^*, \widetilde{p}_g^{*i}) - (p_\pi^*, p_g^{*i})\| \le \varepsilon \text{ for } i = 1, \ldots, n, \\ (\widetilde{p}_\pi^*, \widetilde{p}_g^*) \in \widehat{N}(y^j; \mathrm{cl}\, S^j), & \|(\widetilde{p}_\pi^*, \widetilde{p}_g^*) - (p_\pi^*, p_g^*)\| \le \varepsilon \text{ for } j = 1, \ldots, m, \\ (\widetilde{p}_\pi^*, \widetilde{p}_g^*) \in \widehat{N}(w; \mathrm{cl}\, W), & \|(\widetilde{p}_\pi^*, \widetilde{p}_g^*) - (p_\pi^*, p_g^*)\| \le \varepsilon. \end{cases} \quad (4.21)$$

Applying now the smooth variational descriptions of Fréchet normals in (4.21) from assertions (i) and (ii) Theorem 3.4, we complete the proof of this theorem. △

5. Exact versions of the extended second welfare theorem for economies with public goods

In this section we establish necessary optimality conditions for Pareto and weak Pareto optimal allocations of the nonconvex economy $\mathcal{E}$ with public goods in the *exact/pointwise* form of the extended second welfare theorem under additional SNC assumptions imposed on the sets involved in the description of the economy $\mathcal{E}$. Note that SNC property and its partial modifications seem to be the weakest among compactness-like requirements needed for exact forms of the second welfare theorem. As mentioned in Sect. 3 and fully discussed in [30, Sect. 1.1.4], the basic SNC property is generally weaker than the CEL assumption imposed in the corresponding extensions [12, 17, 18] for economies with only private goods. In this way we get improvements of the second welfare theorem even in the classical settings of convex economies with both private and public goods. We also present a decentralized version of the exact second welfare theorem for nonconvex economies via nonlinear prices.

We begin with the basic version of the second welfare theorem with marginal prices formalized via the M-normal cone (3.2) at Pareto and weak Pareto optimal allocations.

Theorem 5.1 (exact form of the extended second welfare theorem via marginal prices). *Let $(\bar{x}, \bar{y})$ be a Pareto (resp. weak Pareto) optimal allocation of the economy $\mathcal{E}$ with public goods satisfying the corresponding assumptions of Theorem 4.1. Assume in addition that the preference sets $\mathrm{cl}\, P^i(\bar{x})$ are SNC at $\bar{x}^i$ for all $i = 1, \ldots, n$. Then there exist marginal prices $(p_\pi^*, p_g^*) \in E_\pi^* \times E_g^*$ and $p_g^{*i} \in E_g^*$ as $i = 1, \ldots, n$ with*

$$(p_\pi^*, p_g^{*i}) \neq 0 \text{ for at least one } i \in \{1, \ldots, n\} \quad (5.1)$$

satisfying the normal cone inclusions

$$-(p_\pi^*, p_g^{*i}) \in N\big(\bar{x}^i; \mathrm{cl}\, P^i(\bar{x})\big), \quad i = 1, \ldots, n, \tag{5.2}$$

$$(p_\pi^*, p_g^*) \in N(\bar{y}^j; \mathrm{cl}\, S^j), \quad j = 1, \ldots, m, \tag{5.3}$$

$$(p_\pi^*, p_g^*) \in N(\bar{w}; \mathrm{cl}\, W), \tag{5.4}$$

and the underlying equality (4.1) *for the prices associated with public goods.*

Proof. We know from the proof of Theorem 4.1 that $(\bar{x}, \bar{y}, \bar{w})$ is an *extremal point* of the systems of sets $\{\Omega_1, \Omega_2\}$ defined in (4.7) and (4.8) under the NDQ (resp. NDWQ) condition in the case of Pareto (resp. weak Pareto) optimal allocations of the economy $\mathcal{E}$. To get a pointwise version of extended second welfare theorem, we can apply the exact extremal principle from assertion (ii) of Theorem 3.7. In this way we obtain, similarly to the proof of Theorem 4.1, the conclusions of Theorem 5.1 under consideration in the case of finite-dimensional commodity spaces. However, in infinite dimensions this approach requires imposing the SNC assumption on *all* of the sets

$$\mathrm{cl}\, P^i(\bar{x}),\ i = 1, \ldots, n; \quad \mathrm{cl}\, S^j,\ j = 1, \ldots, m; \quad \mathrm{cl}\, W. \tag{5.5}$$

In what follows we do not apply the exact extremal principle directly but pass to the limit from the results of Theorem 4.1 based on the approximate version of the extremal principle. This allows us to arrive at all the conclusions (4.1), (5.1)–(5.4) of Theorem 5.1 under less restrictive SNC assumptions made; see also Remark 5.2 below.

To proceed, take any sequence of $\varepsilon_k \downarrow 0$ as $k \to \infty$ and find by Theorem 4.1 triples

$$(x^k, y^k, w^k) \in \prod_{i=1}^n \mathrm{cl}\, P^i(\bar{x}) \times \prod_{j=1}^m \mathrm{cl}\, S^j \times \mathrm{cl}\, W \ \text{ with } \ (x^k, y^k, w^k) \to (\bar{x}, \bar{y}, \bar{w}) \ \text{ as } \ k \to \infty$$

and prices $(p_\pi^{*k}, p_g^{*k}, p_g^{*ik})$ satisfying all the conclusions of Theorem 4.1 for $\varepsilon = \varepsilon_k$.

It follows from (4.5) that the price sequences $\{(p_\pi^k, p_g^{*ik})\}$ are bounded for all $i = 1, \ldots, n$. Taking into account that the commodity space E is Asplund and hence any bounded subset of E^* is sequentially compact in the weak* topology of E^*, we find (p_π^*, p^{*i}) such that

$$(p_\pi^{*k}, p_g^{*ik}) \xrightarrow{w^*} (p_\pi^*, p_g^{*i}) \ \text{ as } \ k \to \infty, \quad i = 1, \ldots, n, \tag{5.6}$$

along some subsequences, without relabeling. Setting

$$p_g^* := \sum_{i=1}^n p_g^{*i}$$

and passing to the weak* limit in the equality

$$p_g^{*k} := \sum_{i=1}^{n} p_g^{*ik}, \quad k \in I\!N,$$

we get that $p_g^{*k} \rightarrow p_g^*$ as $k \rightarrow \infty$. Passing further to the weak* in the relationships (4.2)–(4.4) with $\varepsilon = \varepsilon_k$ as $k \rightarrow \infty$ and using definition (3.2) of the M-normal cone allow us to conclude that the limiting prices $(p_\pi^*, p_g^*, p_g^{*i})$ satisfy the relationships in (5.2)–(5.4).

It remains to justify the nontriviality condition (5.1) under the SNC requirements imposed in the theorem. To proceed, assume the contrary, i.e.,

$$p_\pi^* = 0 \text{ and } p_g^{*i} = 0 \text{ for all } i = 1, \ldots, n.$$

Then we have by (5.6) that

$$(p_\pi^{*k}, p_g^{*ik}) \xrightarrow{w^*} (0, 0) \text{ as } k \rightarrow \infty, \quad i = 1, \ldots, n.$$

This implies by the SNC assumptions imposed in the theorem that

$$\|(p_\pi^{*k}, p_g^{*ik})\| \rightarrow (0, 0) \text{ as } k \rightarrow \infty, \quad i = 1, \ldots, n. \tag{5.7}$$

It follows from (5.7) and (4.1) that $\|p_g^{*k}\| \rightarrow 0$ as $k \rightarrow \infty$. The latter combined with (5.7) contradicts (4.5) as $\varepsilon = \varepsilon_k$ for large $k \in I\!N$ and thus completes the proof. △

Remark 5.2 (SNC assumptions). It follows from the proof of Theorem 5.1 and Definition 3.6(ii,iii) that the SNC requirements on the preference sets cl $P_i(\bar{x})$ in the theorem can be relaxed to keep the nontriviality condition (5.1) with taking into account the product structure of the commodity space $E = E_\pi \times E_g$. Indeed, it is sufficient to assume that:

- Either *one* of the sets in (5.3) and (5.4) is *strongly PSNC* with respect of the first component while *all* the sets in (5.2) are *PSNC* with respect to the second component at the corresponding points.
- Or *all* the sets in (5.2) are *strongly PSNC* with respect to the second component while *one* of the sets in (5.3) and (5.4) is *PSNC* with respect to the first component at the corresponding points.

Observe that in the case of economies with only private goods both requirements above reduce to imposing the SNC property on *one* of the set in (5.5) at the corresponding point, which is the content of [31, Theorem 8.8].

Next we present two useful specifications of Theorem 5.1 for economies with public goods and additional structural requirements on their initial data. The first one concerns economies with a special structure of the net demand constraint set, which includes the case of *implicit free disposal of commodities*.

Corollary 5.3 (excess demand condition). *Suppose that in the setting of Theorem* 5.1 *the net demand constraint set* $W \subset E_\pi \times E_g$ *admits the representation*

$$W = (\omega_\pi, \omega_g) + \mathrm{cl}\,\Gamma, \tag{5.8}$$

where $(\omega_\pi, \omega_g) \in \mathrm{cl}\,W$ *and* Γ *is a nonempty convex subcone of* $E_\pi \times E_g$. *There are marginal prices* $(p_\pi^*, p_g^*) \in E_\pi^* \times E_g^*$ *and* $p_g^{*i} \in E_g^*$ *as* $i = 1, \ldots, n$ *satisfying all the relationships in* (4.1), (5.1)–(5.3) *and such that*

$$\Big\langle (p_\pi^*, p_g^*), \sum_{i=1}^{n} (\bar{x}_\pi^i, \bar{x}_g^i) - \sum_{j=1}^{m} \bar{y}^j - (\omega_\pi, \omega_g) \Big\rangle = 0. \tag{5.9}$$

Proof. By Theorem 5.1 it remains to show that inclusion (5.4) for the special conic structure of W in (5.8) implies the *zero value of excess demand* condition (5.9) at marginal prices, which is an economic manifestation of the *complementary slackness* condition in optimization. To proceed, observe from relationships (5.4), (5.8) and the normal cone representation (3.6) for convex sets that

$$\big\langle (p_\pi^*, p_g^*), \bar{w} - (\omega_\pi, \omega_g) \big\rangle \geq \big\langle (p_\pi^*, p_g^*), w - (\omega_\pi, \omega_g) \big\rangle \;\text{ for all }\; w \in \mathrm{cl}\,W. \tag{5.10}$$

Hence $\langle (p_\pi^*, p_g^*), (\bar{w} - (\omega_\pi, \omega_g) \rangle \geq 0$. On the other hand, we have

$$2\big(\bar{w} - (\omega_\pi, \omega_g)\big) \in W - (\omega_\pi, \omega_g) = \Gamma$$

due to the conic structure of Γ, which implies by (5.10) that $\langle (p_\pi^*, p_g^*), \bar{w} - (\omega_\pi, \omega_g) \rangle \leq 0$. This yields (5.9) and completes the proof of the corollary. $\triangle$

In the case of economies with *convex* preference and production sets considered in the next corollary of Theorem 5.1, relationships (5.2) and 5.3 – unified with those in (4.1) and (5.2) for prices corresponding to public goods – reduce to the classical *consumer expenditure minimization* and *firm profit maximization* conditions of the second fundamental theorem of welfare economics. We are able, however, to significantly improve the classical *interiority condition* required for the validity of the second welfare theorem with infinite commodities in convex settings. As known from [30, Theorem 1.17], the SNC property imposed in our Theorem 5.1 is equivalent, in the case of convex sets with nonempty relative interiors, to their *finite codimension*. Furthermore, convex sets in Asplund spaces may be SNC even having empty relative interiors; see [30, Example 3.6]. The PSNC extensions discussed above in Remark 5.2 signify further far-going departures from the interiority condition.

Corollary 5.4 (second welfare theorem in convex settings with no interiority requirements). *In addition to the assumptions of Theorem* 5.1, *suppose* *that the sets* $\operatorname{cl} P^i(\bar{x})$ *as* $i = 1, \ldots, n$ *and* $\operatorname{cl} S^j$ *as* $j = 1, \ldots, m$ *are convex. Then there are prices* $(p_\pi^*, p_g^*) \in E_\pi^* \times E_g^*$ *and* $p_g^{*i} \in E_g^*$ *as* $i = 1, \ldots, n$ *satisfying the relationships in* (4.1), (5.1), (5.4), *and:*

$$\bar{x}^i \text{ minimizes } \langle (p_\pi^*, p_g^{*i}), x^i \rangle \text{ over } x^i \in \operatorname{cl} P^i(\bar{x}), \quad i = 1, \ldots, n; \tag{5.11}$$

$$\bar{y}^j \text{ maximizes } \langle (p_\pi^*, p_g^*), y^j \rangle \text{ over } y^j \in \operatorname{cl} S^j, \quad j = 1, \ldots, m. \tag{5.12}$$

Proof. Follows from Theorem 5.1 due to the normal cone representation (3.6). △

Let us next present a *decentralized* counterpart of Theorem 5.1 via *nonlinear prices.*

Theorem 5.5 (decentralized version of the extended second welfare theorem for nonconvex economies with public goods). *Let* $(\bar{x}, \bar{y})$ *be a Pareto (resp. weak Pareto) optimal allocation of the economy* $\mathcal{E}$ *under the corresponding assumptions of Theorem* 5.1 *(or the relaxed PSNC assumptions in Remark* 5.2*), and let* $(p_\pi^*, p_g^*) \in E_\pi^* \times E_g^*$ *and* $p_g^{*i} \in E_g^*$ *as* $i = 1, \ldots, n$ *be marginal prices satisfying the conditions in* (4.1) *and* (5.1)–(5.4)*. Then the following assertions hold:*

(i) *There exist sequences of nonlinear prices* $f_k = (f_k^1, \ldots, f_k^n)$ *and* $h_k = (h_k^1, \ldots, h_k^{m+1})$ *as well as sequences of suboptimal allocations*

$$(x_k, y_k) \in \prod_{i=1}^n \operatorname{cl} P^i(\bar{x}) \times \prod_{j=1}^m \operatorname{cl} S^j \quad \text{with} \quad w_k := \sum_{i=1}^n x_k^i - \sum_{j=1}^m y_k^j \in \operatorname{cl} W$$

such that for all $k \in I\!N$ *we have the decentralized relationships:*

- *Each* f_k^i *and* h_k^j *is Fréchet differentiable at* x_k^i, y_k^j, *and* w_k *achieving its global minimum over* $\operatorname{cl} P^i(\bar{x})$ *and its global maximum over* $\operatorname{cl} S^j$ *and* $\operatorname{cl} W$ *at these points for* $i = 1, \ldots, n$, $j = 1, \ldots, m$, *and* $j = m + 1$, *respectively.*
- $(x_k, y_k, w_k) \to (\bar{x}, \bar{y}, \bar{w})$ *as* $k \to \infty$ *with the equilibrium price convergence*

$$\begin{cases} (p_{\pi k}^*, p_{gk}^{*i}) := \nabla f_k^i(x_k^i) \stackrel{w^*}{\to} (p_\pi^*, p^{*i}), \quad i = 1, \ldots, n, \\[2ex] (p_{\pi k}^{*j}, p_{gk}^{*j}) := \nabla h_k^j(y_k^j) \stackrel{w^*}{\to} (p_\pi^*, p_g^*), \quad j = 1, \ldots, m, \\[2ex] (p_{\pi k}^{*m+1}, p_{gk}^{*m+1}) := \nabla h_k^{m+1}(w_k) \stackrel{w^*}{\to} (p_\pi^*, p_g^*) \text{ as } k \to \infty. \end{cases}$$

(ii) *If in addition to the assumptions in* (i) *the commodity space E is Fréchet smooth, then the nonlinear prices f_k^i and h_k^j can be chosen to be Fréchet differentiable on E and such that each f_k^i, $i = 1, \ldots, n$, is convex and achieves its global minimum over* $\operatorname{cl} P^i(\bar{x})$ *uniquely at x^i while each h_k^j, $j = 1, \ldots, m + 1$, is concave and achieves its global maximum over* $\operatorname{cl} S^j$ *for $j = 1, \ldots, m$ and over* $\operatorname{cl} W$ *for $j = m + 1$ uniquely at y^j and w, respectively.*

Proof. It follows by using the arguments similar to those in the proof of Theorem 5.1 with applying there Theorem 4.4 at each approximation step $k \in I\!N$ and taking into account the strong convergence of the derivatives in (4.20) for $\varepsilon = \varepsilon_k \downarrow 0$ as $k \to \infty$. △

We conclude this section with some discussions on possible developments and generalizations of the results obtained for nonconvex economies with public goods.

Remark 5.6 (further developments). Similarly to recent developments for economies with only private goods, we have the following modifications and generalizations of the above versions of the extended second welfare theorem for nonconvex economies with public goods:

(i) Counterparts of the results above hold for *strong Pareto* optimal allocations and for the new notion of *strict Pareto* ones [4] under the corresponding modifications of the net demand qualification conditions defined in [4] for economies with only private goods.

(ii) In [16] some refinement of the NDQ condition for Pareto optimal allocation was introduced and employed to the second welfare theorem for economies with only private goods. Analogs of this condition for strong and strict Pareto optimal allocations and the corresponding versions of the second welfare theorem were given in [5] for private goods economies. Following the scheme in [5], we can extend these conditions and results to nonconvex economies with public goods.

6. Nonconvex economies with ordered commodity space

This concluding section of the paper concerns nonconvex economies with public goods and *ordered* infinite-dimensional spaces of commodities. First we specify the results obtained above for Pareto and weak Pareto optimal allocations and then establish their new counterparts for the case of strong Pareto optimal allocations.

Let E be an *ordered Banach space* with the *closed positive cone*

$$E_+ := \{e \in E \mid e \geq 0\},$$

where the standard partial ordering relation is denoted by "≥' in accordance to the conventional notation in the economic literature. The associated *dual closed positive cone* E^*_+, which is the closed positive cone of the ordered space E^*, admits the representation

$$E^*_+ := \{e^* \in E^* \mid e^* \geq 0\} = \{e^* \in E^* \mid \langle e^*, e\rangle \geq 0 \text{ for all } e \in E_+\},$$

where the order on E^* is induced by the given one "≥" on E.

The next theorem provides efficient conditions ensuring the *positivity* of the marginal prices associated with both private and public goods in our extended second welfare theorem. The result is given for weak Pareto optimal allocations, and hence it holds for any stronger notions of Pareto optimality in the welfare economic model under consideration.

Theorem 6.1 (price positivity in the extended second welfare theorem for ordered commodities). *Let $(\bar{x}, \bar{y})$ be a weak Pareto optimal allocation of the model $\mathcal{E}$ with public goods, where $E = E_\pi \times E_g$ is an ordered commodity space. Suppose the fulfillment of all the assumptions of Theorem* 5.1 *but the NDWQ condition and assume instead that each consumer $i \in \{1, \ldots, n\}$ satisfies the following desirability condition:*

$$\operatorname{cl} P^i(\bar{x}) + E_+ \subset P^i(\bar{x}). \tag{6.1}$$

Then there exist positive marginal prices $(p^*_\pi, p^*_g) \in E^*_\pi \times E^*_g$ and $p^{*i}_g \in E^*_g$ as $i = 1, \ldots, n$ for which all the conclusions (4.1), (5.1)–(5.4) *hold.*

Proof. First observe that the desirability condition (6.1) implies the NDWQ condition by Proposition 2.4(ii). Furthermore, by [31, Lemma 8.12] we have the implication

$$[\Omega - E_+ \subset \Omega] \Longrightarrow [N(\bar{e}; \Omega) \subset E^*_+] \tag{6.2}$$

for an arbitrary closed subset $\Omega \subset E$ of an ordered Banach space and any $\bar{e} \in \Omega$. Thus all the conclusions of Theorem 5.1 holds for the weak Pareto optimal allocation $(\bar{x}, \bar{y})$ of the economy $\mathcal{E}$. Then it follows from (5.2) that $p^*_\pi \geq 0$ and $p^{*i}_g \geq 0$ for all $i = 1, \ldots, n$. This yields $p^*_g \geq 0$ by (4.1) and completes the proof of the theorem. △

Remark 6.2 (more on price positivity under free disposal or implicit free disposal of commodities). Consider the following conditions of the free disposal type:

- There is $j \in \{1, \ldots, m\}$ such that the jth firm satisfies the *free disposal condition*

$$\text{cl}\, S^j - E_+ \subset \text{cl}\, S^j. \tag{6.3}$$

- The net demand constraint set W exhibits the *implicit free disposal* of commodities

$$\text{cl}\, W - E_+ \subset \text{cl}\, W. \tag{6.4}$$

Then it follows from (5.3), (5.4), and (6.2)–(6.4) that we have $(p_\pi^*, p_g^*) \geq 0$ provided that either condition (6.3) holds for some $j \in \{1, \ldots, m\}$, or condition (6.4) is satisfied. In particular (unifying the conclusions of Theorem 6.1 and this remark), for the case of economies with only private goods we have that $p_\pi^* \geq 0$ if either one of the consumers $i \in \{1, \ldots, n\}$ satisfies the desirability condition (6.1), or one of the firms $j \in \{1, \ldots, m\}$ satisfies the free disposal condition (6.3), or the implicit free disposal condition (6.4) holds.

Next we derive refined versions of the extended second welfare theorem for strong Pareto optimal allocations of economies with public goods. In contrast to the corresponding results of Theorem 5.1 for Pareto and weak Pareto optimal allocations, the NDQ and NDWQ conditions may not be satisfied. Recall that the closed positive cone $E_+ \subset E$ is *generating* for E if $E = E_+ - E_+$. The class of normed spaces ordered by their generating positive cones is sufficiently large including, in particular, all Banach lattice (or complete Riesz spaces); see, e.g., [12, 24] and the references therein.

Theorem 6.3 (extended second welfare theorem for strong Pareto optimal allocations). *Let $(\bar{x}, \bar{y})$ be a strong Pareto optimal allocation of the economy $\mathcal{E}$ with public goods, and let all the assumptions of Theorem* 5.1, *except the NDQ/NDWQ conditions, be satisfied. Suppose instead that the positive cone E_+ of the commodity space $E = E_\pi \times E_g$ is generating, that the sets S^j and W are locally closed around $\bar{y}^j$ and $\bar{w}$, respectively, and that one of the following conditions holds:*

- *The free disposal of commodities*

$$S^j - E_+ \subset S^j \;\; \textit{for some} \;\; j \in \big\{1, \ldots, m\big\}.$$

- *The implicit free disposal of commodities* (6.4).
- $n > 1$, *and there is $i_0 \in \{1, \ldots, n\}$ with $\text{cl}\, P^{i_0}(\bar{x}) \neq \emptyset$ such that*

$$\text{cl}\, P^i(\bar{x}) + E_+ \subset \text{cl}\, P^i(\bar{x}) \;\; \textit{for some} \;\; i \in \big\{1, \ldots, n\big\} \setminus \big\{i_0\big\}.$$

Then there exist marginal prices $(p_\pi^*, p_g^*) \in E_\pi^* \times E_g^*$ *and* $p_g^{*i} \in E_g^*$ *as* $i = 1, \ldots, n$ *satisfying conditions* (4.1) *and* (5.1)–(5.4)*. If in addition assumption* (6.1) *holds for all* $i \in \{1, \ldots, n\}$*, then we have the price positivity*

$$p^{*i} \geq 0 \text{ for all } i = 1, \ldots, n \text{ and } (p_\pi^*, p_g^*) \geq 0. \tag{6.5}$$

Proof. For definiteness, consider only the case of the implicit free disposal of commodities (6.4); the other two cases of the theorem are treated similarly. Observe first that

$$\bar{w} = \sum_{i=1}^{n} \left(\bar{x}_\pi^i, \bar{x}_g\right) - \sum_{j=1}^{m} \bar{y}^j$$

is a *boundary point* of the net demand constraint set W. Indeed, assuming the contrary gives us a point $\bar{e} = (\bar{e}_\pi, \bar{e}_g) \neq 0$ such that $\bar{w} + \bar{e} \in W$, which implies $(\bar{x} + \bar{e}, \bar{y})$ is a feasible allocation of the economy $\mathcal{E}$. This contradicts the strong Pareto optimality of $(\bar{x}, \bar{y})$. Since $\bar{w}$ is a boundary point of W, there is a sequence $\{e^k\} \subset E$ converging to zero and such that

$$\bar{w} + e^k \notin W \text{ for all } k \in I\!N. \tag{6.6}$$

Taking into account that the cone E_+ is generating and employing the the classical Krein–Šmulian theorem from [1], we find a constant $M > 0$ such that for every $e \in E$ there exist vectors $u, v \in E_+$ satisfying the conditions

$$e = u - v \text{ and } \max\left\{\|u\|, \|v\|\right\} \leq M\|e\|.$$

The latter and the relationship in (6.6) apply the existence of sequences $\{u^k\} \subset E_+$ and $\{v^k\} \subset E_+$ with $e^k = u^k - v^k$ and

$$\bar{w} + u^k \notin W \text{ for all } k \in I\!N \text{ and } u^k \overset{E_+}{\to} 0 \text{ as } k \to \infty. \tag{6.7}$$

Consider now the sets Ω_1 and Ω_2 defined in (4.7) and (4.8), respectively, where the closure operation for S^j and W can be omitted. Let us show that $(\bar{x}, \bar{y}, \bar{w}) \in \Omega_1 \cap \Omega_2$ is an *extremal point* of the system $\{\Omega_1, \Omega_2\}$. We need to find a sequence $\{a^k\} \subset E^{n+m+1}$ with $a^k \to 0$ as $k \to \infty$ such that the extremality condition (4.9) is satisfied. To proceed, define $a^k := (0, \ldots, u^k)$ and suppose that (4.9) does not hold along this sequence $\{a^k\}$, i.e., there are $(x^k, y^k, w^k) \in \Omega_1$ such that $(x^k, y^k, w^k - u^k) \in \Omega_2$ for all $k \in I\!N$. This yields that

$$\sum_{i=1}^{n} \left(x_\pi^{ki}, x_g^k\right) - \sum_{j=1}^{m} y^{kj} = w^k - u^k \in W,$$

which implies that (x^k, y^k) is a feasible allocation of the economy $\mathcal{E}$. By the *strong Pareto* optimality of $(\bar{x}, \bar{y})$ we get that $(x^k, y^k) = (\bar{x}, \bar{y})$ for all $k \in I\!N$ and hence

$$
\begin{aligned}
\bar{w} + u^k &= \sum_{i=1}^{n} \left(\bar{x}_\pi^i, \bar{x}_g\right) - \sum_{j=1}^{m} \bar{y}^j + u^k \\
&= \sum_{i=1}^{n} \left(x_\pi^{ki}, x_g^k\right) - \sum_{j=1}^{m} y^{kj} + u^k \\
&= (w^k - u^k) + u^k = w^k \in W, \quad k \in I\!N,
\end{aligned}
$$

which contradicts (6.7) and thus justifies the extremality of $(\bar{x}, \bar{y}, \bar{w})$. Applying now the *extremal principle* from Theorem 3.7(i) to the system $\{\Omega_1, \Omega_2\}$ at $(\bar{x}, \bar{y}, \bar{w})$ and arguing as in the proof of Theorem 4.1, we obtain approximate marginal prices satisfying all the conclusions of the latter theorem with $\varepsilon = \varepsilon_k \downarrow 0$ as $k \to \infty$. Passing then to the limit as $k \to \infty$ as in the prove of Theorem 5.1 under the SNC assumptions made allows us to arrive at the marginal prices $(p_\pi^*, p_g^*, p_g^{*i})$ satisfying conditions (4.1) and (5.1)–(5.4). Finally, the price positivity (6.5) under the additional assumption (6.1) follows from Theorem 6.1. △

References

1. Abramovich, Y.A., Aliprantis, C.D.: An Invitation to Operator Theory. American Mathematical Society, Providence, RI (2002)
2. Alpirantis, C.D., Tourky, R., Yannelis, N.C.: A theory of values with nonlinear prices. Equilibrium analysis beyond vector lattices. J. Econ. Theory **100**, 22–72 (2001)
3. Arrow, K.J.: An extension of the basic theorem of classical welfare economics. In: Neyman, J. (ed.) Proceedings of the Second Berkeley Symposium on Mathematical Statistics and Probability, pp 507–532. University of California, Berkeley, CA (1951)
4. Bao, T.Q., Mordukhovich, B.S.: Set-valued optimization in welfare economics. Adv. Math. Econ. **13**, 113–153 (2010)
5. Bao, T.Q., Mordukhovich, B.S.: Refined necessary conditions in multiobjective optimization with applications to microeconomic modeling. Commun. Pure Appl. Anal. (2010, to appear)
6. Bonnisseau, J.M., Cornet, B.: Valuation equilibrium and Pareto optimum. J. Math. Econ. **17**, 293–308 (1988)
7. Bonnisseau, J.M., Lachiri, O.: About the second theorem of welfare economics with stock markets. Pac. J. Math. **2**, 469–485 (2006)
8. Borwein, J.M., Strójwas, H.M.: Tangential approximations. Nonlinear Anal. **9**, 1347–1366 (1985)

9. Cornet, B.: The second welfare theorem in nonconvex economics. CORE discussion paper, no 8630 (1986)
10. Debreu, G.: The coefficient of resource utilization. Econometrica **19**, 273–292 (1951)
11. Debreu, G., Hildenbrand, W.: Existence of equilbria in market with infinite-dimensional commodity space. Technical report, University of California Berkeley, CA (1968)
12. Florenzano, M., Gourdel, P., Jofré, A.: Supporting weakly Pareto optimal allocations in infinite-dimensional nonconvex economies. J. Econ. Theory **29**, 549–564 (2006)
13. Foley, D.K.: Lindal's solution and the core of an economic with public goods. Econometrica **38**, 66–72 (1967)
14. Guesnerie, R.: Pareto optimality in nonconvex economies. Econometrica **43**, 1–29 (1975)
15. Hicks, J.R.: The foundations of welfare economics. Econ. J. **49**, 696–712 (1939)
16. Ioffe, A.D.: Variational analysis and mathematical economics, I: subdifferential calculus and the second theorem of welfare economics. Adv. Math. Econ. **12**, 71–95 (2009)
17. Jofré, A.: A second welfare theorem in nonconvex economies. In: Théra (ed.) Constructive, Experimental and Nonlinear Analysis. CMS Conference Proceedings, vol. 27, pp. 123–164 (2000)
18. Jofré, A., Rivera Cayopi, J.: A nonconvex separation property and some applications. Math. Program **108**, 37–51 (2006)
19. Khan, M.A.: Ioffe's normal cone and the foundation of welfare economics: the infinite dimensional theory. J. Math. Anal. Appl. **161**, 284–298 (1991)
20. Khan, M.A.: The Mordukhovich normal cone and the foundations of welfare economics. J. Public Econ. Theory 1:309–338 (1999)
21. Khan, M.A., Vohra, R.: An extension of the second welfare theorem to economies with nonconvexity and public goods. Q. J. Econ. **102**, 223–245 (1987)
22. Lang, O.: The foundation of welfare economics. Econometrica **10**, 215–228 (1942)
23. Malcolm, G.G., Mordukhovich, B.S.: Pareto optimality in nonconvex economies with infinite-dimensional commodity spaces. J. Global Optim. **20**, 323–346 (2001)
24. Mas-Colell, A.: The price equilibrium existence problem in topological vector lattices. Econometrica **54**, 1039–1053 (1986)
25. Milleron, J.C.: Theory of value with public goods: a survey article. J. Econ. Theory **5**, 419–477 (1972)

26. Mordukhovich, B.S.: Maximum principle in time optimal control problems with nonsmooth constraints. J. Appl. Math. Mechan. **40**, 960–969 (1976)
27. Mordukhovich, B.S.: Metric approximations and necessary optimality conditions for general classes of nonsmooth extremal problems. Soviet Math. Dokl. **22**, 526–530 (1980)
28. Mordukhovich, B.S.: Abstract extremal principle with applications to welfare economics. J. Math. Anal. Appl. **251**, 187–216 (2000)
29. Mordukhovich, B.S.: Nonlinear prices in nonconvex economies with classical Pareto and strong Pareto optimal allocations. Positivity **9**, 541–568 (2005)
30. Mordukhovich, B.S.: Variational analysis and generalized differentiation, I: basic theory. In: Grundlehren Series (Fundamental Principles of Mathematical Sciences), vol. 330. Springer, Berlin (2006)
31. Mordukhovich, B.S.: Variational analysis and generalized differentiation, II: applications. In: Grundlehren Series (Fundamental Principles of Mathematical Sciences), vol. 331. Springer, Berlin (2006)
32. Rockafellar, R.T.: Directional Lipschitzian functions and subdifferential calculus. Proc. Lond. Math. Soc. **39**, 331–355 (1979)
33. Samuelson, P.A.: Further commentary on welfare economics. Am. Econ. Rev. **33**, 605–607 (1943)
34. Samuelson, P.A.: Foundation of Economic Analysis. Harvard University Press, Cambridge, MA (1947)
35. Samuelson, P.A.: The pure theory of public expenditure. Rev. Econ. Stat. **36**, 387–389 (1954)
36. Simone, A.D., Graziano, M.G.: The pure theory of public goods: the case of many commodities. J. Math. Econ. **40**, 847–868 (2004)
37. Villar, A.: Equilibrium and Efficiency in Production Economies. Springer, Berlin (2000)

Adv. Math. Econ. 14, 127–145 (2011)

Advances in
MATHEMATICAL
ECONOMICS

Partial hedging for defaultable claims

Yumiharu Nakano[1,2]

[1] Graduate School of Innovation Management, Tokyo Institute of Technology, 2-12-1 Ookayama, Tokyo 152-8552, Japan

[2] PRESTO, Japan Science and Technology Agency, 4-1-8 Honcho Kawaguchi, Saitama 332-0012, Japan
(e-mail: yumiharu.nakano@gmail.com)

Received: February 22, 2010
Revised: May 25, 2010

JEL classification: C61, G11

Mathematics Subject Classification (2010): Primary 91B28; Secondary 60G55

Abstract. The subject of this paper is the problems of finding optimal hedging portfolios for defaultable claims in incomplete markets modeled by Itô processes, in the case where the portfolio processes are adapted to the full filtration. Two kinds of optimizations are considered: the maximization of the probability of super-hedge and the minimization of the expected discounted loss of hedging. We combine a super-hedging argument with Neyman–Pearson lemma in the hypothesis testing to reduce the original dynamic problems to static ones. The convex duality method as in Cvitanić and Karatzas (Bernoulli 7:79–97, 2001) plays a key role to solve the reduced problems.

Key words: hedging, defaultable claims, convex duality, Neyman–Pearson lemma, jump processes

1. Introduction

It is known that, in arbitrage-free, incomplete financial markets, the super-hedging cost of a contingent claim is often too high. More precisely, for any European call option in markets with transaction costs, the cheapest super-hedging is given by the buy-and-hold portfolio. This result is conjectured by Davis and Clark [11], and proved by, to name a few, Soner, Shreve and Cvitanić [28], Cvitanić, Pham and Touzi [9], Levental and Skorohod [17],

DOI: 10.1007/978-4-431-53883-7_6,

and Jakubenas, Levental and Ryznar [15]. Similar results are obtained by Bellamy–Jeanblanc [2] in jump-diffusion models and by Cvitanić, Pham and Touzi [10] in stochastic volatility models.

In such a situation, it is reasonable that a hedger of a claim starts with an initial capital less than the super-hedging cost and accepts the possibility of the shortfall. One criterion for measuring this downside risk is the probability of super-hedging being successful. Optimizing this criterion is usually called quantile hedging, which is first studied by Kulldorff [16] in the context of gambling theory. Browne [5] considers the case of financial markets modeled by Itô processes with deterministic coefficients. Föllmer and Leukert [12] studies this problem for general semimartingale financial market models. Spivak and Cvitanić [29] treats partial information market models and markets with different interest rates for borrowing and for lending. Sekine [26] analyzes the case of defaultable claims in the Brownian market models.

Taking the shortfall size into consideration, Föllmer and Leukert [13] and Cvitanić [6] study the minimization problem of the expected discounted shortfall. Other criterions, such as the expected (strictly convex) loss function or the risk measures, for the shortfall risk are also considered. See Cvitanić and Karatzas [7], Föllmer and Leukert [13], Nakano [18–20], Pham [23], Rudloff [24, 25], and Sekine [27], for examples.

In this paper, we first consider the quantile hedging problem for defaultable claims in Brownian market models as in [27]. It investigates the case where the portfolios are adapted to the market information structure and gives closed form solutions by some reductions of the original problems to default-free ones. In our framework presented below, the portfolio processes are assumed to be adapted to the full filtration, i.e., the filtration generated by both the price and default indicator processes. We also consider the minimization of the expected discounted loss of the shortfall in the same framework.

To solve these problems, following [12] and [13], we combine a super-hedging argument with a Neyman–Pearson lemma in the hypothesis testing to reduce the original dynamic problems to static ones. In a complete market framework, the reduced static problems can be stated as the testing problems of a single null hypothesis versus a single alternative hypothesis, and so is directly solved by the classical Neyman–Pearson lemma. However, this is not the case in our incomplete markets. To handle this issue, as in [6] and [19], we follow the convex duality approach for the generalized Neyman–Pearson lemma developed in [8].

This paper is organized as follows: In Sect. 2, we describe our market models. As a basic result, we give an explicit formula for the super-replication cost. Section 3 presents a solution to our quantile hedging problem for defaultable claims with zero recovery rate. In doing so, we explicitly solve the dual problem with the help of a good structure of the class of the equivalent martingale measures. In Sect. 4, we solve the minimization problem of the expected discounted shortfall using the argument similar to that in Sect. 3. Section 5 deals with the case of non-zero recovery rate. Section 6 concludes this paper.

2. Model

We consider the financial market with terminal time $T \in (0, \infty)$ consisting of one stock with price process $\{S_t\}_{0 \le t \le T}$ and one riskless bond with price process $\{B_t\}_{0 \le t \le T}$, whose dynamics are given respectively by

$$\begin{aligned} dS_t &= S_t\{b_t dt + \sigma_t dW_t\}, \quad 0 \le t \le T, \quad S_0 = s_0 \in (0, \infty), \\ dB_t &= r_t B_t dt, \quad 0 \le t \le T, \quad B_0 = 1. \end{aligned}$$

Here, $\{W_t\}_{t \ge 0}$ is a standard one-dimensional Brownian motion on a complete probability space $(\Omega, \mathcal{G}, \mathbb{P})$. The filtration $\mathbb{F} = \{\mathcal{F}_t\}_{t \ge 0}$ is generated by $\{W_t\}_{t \ge 0}$, augmented with $\mathbb{P}$-null sets in $\mathcal{G}$. The processes $\{b_t\}$, $\{r_t\}$, and $\{\sigma_t\}$ are all assumed to be bounded $\mathbb{F}$-predictable processes. Moreover we assume that $\sigma_t > 0$ for $t \in [0, T]$ a.s. and that $\{\sigma_t^{-1}\}$ is also bounded. Then, the process

$$\theta_t := \sigma_t^{-1}(b_t - r_t), \quad 0 \le t \le T$$

is a bounded $\mathbb{F}$-predictable process.

Let τ be a positive random variable satisfying $\mathbb{P}(\tau > t) > 0$ for any $t \ge 0$, and let $\{N_t\}_{t \ge 0}$ be the counting process with respect to τ, i.e.,

$$N_t = 1_{\{\tau \le t\}}, \quad t \ge 0.$$

Denote by $\mathbb{H} = \{\mathcal{H}_t\}_{t \ge 0}$ the filtration generated by $\{N_t\}$ and by $\mathbb{G} = \{\mathcal{G}_t\}_{t \ge 0}$ the filtration $\mathbb{F} \vee \mathbb{H}$. For simplicity we assume that $\mathcal{G} = \mathcal{G}_T$. The survival process $\{G_t\}_{t \ge 0}$ of τ with respect to $\mathbb{F}$ is then defined by

$$G_t = \mathbb{P}(\tau > t \mid \mathcal{F}_t), \quad 0 \le t \le T.$$

We assume that $G_t > 0$ for $t \ge 0$, and consider the hazard process $\{\Gamma_t\}_{t \ge 0}$ of τ with respect to $\mathbb{F}$ defined by $G_t = e^{-\Gamma_t}$ or $\Gamma_t = -\log G_t$ for every $t \ge 0$. We also assume that $\Gamma_t = \int_0^t \mu_s ds$, $t \ge 0$, for some nonnegative $\mathbb{F}$-predictable process $\{\mu_t\}_{t \ge 0}$, so-called $\mathbb{F}$-intensity of the random time τ. Then the process

$$M_t := N_t - \int_0^t \mu_s(1 - N_{s-})ds = N_t - \int_0^{t \wedge \tau} \mu_s ds, \quad t \ge 0,$$

follows a $\mathbb{G}$-martingale (see Bielecki and Rutkowski [3]).

We now make the standing assumption that $\{W_t\}$ is a $(\mathbb{G}, \mathbb{P})$-standard Brownian motion. Notice that if τ is independent of $\{W_t\}$ then this assumption is satisfied. Also, for a given $\mathbb{F}$-hazard process, we can construct the random time τ such that $\{W_t\}$ is a $(\mathbb{G}, \mathbb{P})$-standard Brownian motion (see, e.g., [3]).

As in the usual Brownian market models, we consider the $\mathbb{G}$-martingale

$$Z_t^* = \exp\left(-\int_0^t \theta_s dW_s - \frac{1}{2}\int_0^t \theta_s^2 ds\right), \quad 0 \le t \le T.$$

Then by Girsanov's theorem, the process

$$W_t^* := W_t + \int_0^t \theta_s ds, \quad 0 \le t \le T$$

is a standard Brownian motion under the probability measure $\mathbb{P}^*$ defined by

$$\frac{d\mathbb{P}^*}{d\mathbb{P}} = Z_T^*.$$

In addition, we consider the process

$$Z_t^\kappa = (1 + \kappa_\tau 1_{\{\tau \le t\}}) \exp\left(-\int_0^{t\wedge\tau} \kappa_s \mu_s ds\right), \quad 0 \le t \le T,$$

where $\{\kappa_t\}_{0\le t\le T}$ is taken from the class

$$\mathcal{D} = \{\{\kappa_t\}_{0\le t\le T} : \text{bounded, } \mathbb{G}\text{-predictable, } \kappa_t > -1 \ dt \times d\mathbb{P}\text{-a.e.}\}.$$

Then $\{Z_t^\kappa\}$, $\kappa \in \mathcal{D}$, satisfies

$$Z_t^\kappa = 1 + \int_0^t \kappa_s Z_{s-}^\kappa dM_s, \quad 0 \le t \le T,$$

and follows a $(\mathbb{G}, \mathbb{P})$-martingale (see Brémaud [4] for example). Since the quadratic covariation process $[Z^*, Z^\kappa]$ is identically zero,

$$dZ_t^* Z_t^\kappa = Z_t^* Z_{t-}^\kappa(-\theta_t dW_t + \kappa_t dM_t). \tag{2.1}$$

Thus, $\{Z_t^* Z_t^\kappa\}$ is a $(\mathbb{G}, \mathbb{P})$-positive martingale for $\kappa \in \mathcal{D}$. Each $\{Z_t^\kappa\}$ is orthogonal to $(\mathbb{F}, \mathbb{P})$-martingales, so we can show that $\{W_t^*\}$ is also a Brownian motion under $\mathbb{Q}^\kappa$ defined by $d\mathbb{Q}^\kappa/d\mathbb{P} = Z_T^* Z_T^\kappa$. Hence $\{\mathbb{Q}^\kappa : \kappa \in \mathcal{D}\}$ defines the class of the equivalent martingale measures. We refer to [3, Sect. 5.3] for details.

We consider $\mathbb{G}$ as the available information for the market participants. The portfolio process is thus defined as a $\mathbb{G}$-predictable process $\{\pi_t\}_{0\le t\le T}$

satisfying $\int_0^T |\pi_t|^2 dt < \infty$, a.s. The (self-financing) wealth process $\{X_t^{x,\pi}\}_{0\le t\le T}$ for an initial wealth $x \ge 0$ and a portfolio process $\{\pi_t\}$ is then described by

$$dX_t^{x,\pi} = r_t X_t^{x,\pi} dt + \pi_t(b_t - r_t)dt + \pi_t \sigma_t dW_t, \quad X_0^{x,\pi} = x.$$

The solution to this equation is given by

$$X_t^{x,\pi} = B_t \left[x + \int_0^t B_u^{-1} \pi_u \{(b_u - r_u)du + \sigma_u dW_u\} \right], \quad 0 \le t \le T.$$

We write $\mathcal{A}(x)$ for the set of all portfolio processes $\{\pi_t\}_{0\le t\le T}$ such that $X_t^{x,\pi} \ge 0, 0 \le t \le T$, a.s.

By Itô formula and (2.1), we get, for $\pi \in \mathcal{A}(x)$,

$$dL_t^\kappa X_t^{x,\pi} = L_{t-}^\kappa[(\pi_t\sigma_t - X_t^{x,\pi}\theta_t)dW_t + X_t^{x,\pi}\kappa_t dM_t],$$

where

$$L_t^\kappa = B_t^{-1} Z_t^* Z_t^\kappa, \quad \kappa \in \mathcal{D}.$$

This and the nonnegativity of the wealth process mean that the process $\{L_t^\kappa X_t^{x,\pi}\}$ is a supermartingale for each $\pi \in \mathcal{A}(x)$. We denote by $\mathcal{L}$ the set of all random variable L_T^κ, $\kappa \in \mathcal{D}$.

In this setting, we consider hedging problems for the defaultable claim H defined by

$$H = Y 1_{\{\tau > T\}} + \delta Y 1_{\{\tau \le T\}}. \tag{2.2}$$

Here, Y is an $\mathcal{F}_T$-measurable nonnegative random variable, which represents the payoff received by the holder at time T if the default does not occur in $[0, T]$. We assume that $\mathbb{E}^*[Y] < \infty$, where $\mathbb{E}^*$ stands for the expectation with respect to $\mathbb{P}^*$. The constant $\delta \in [0, 1]$ is the recovery rate of the payoff in case the default occurs in $[0, T]$.

The most conservative way of hedging the claims is the so-called super-hedging, and its cost $\Pi(H)$ of H is defined by

$$\Pi(H) = \inf\{x \ge 0 : X_T^{x,\pi} \ge H \text{ a.s. for some } \pi \in \mathcal{A}(x)\}.$$

In our setting, this super-hedging cost can be obtained explicitly.

Proposition 2.1. *Let H be as in* (2.2) *such that $\mathbb{E}^*[Y] < \infty$. Then we have*

$$\Pi(H) = \mathbb{E}^*[B_T^{-1} Y].$$

Moreover, the replicating portfolio for Y becomes a super-hedging portfolio for H.

It should be stressed that the super-hedging portfolio in this proposition is *minimal*. That is, it has the minimal initial cost for hedging.

Proposition 2.1 is showed in [21], but for the reader's convenience we present the full proof.

Proof. Set $\tilde{x} = \mathbb{E}^*[B_T^{-1}Y]$ and let $\tilde{\pi}$ be the replicating portfolio for Y. Then we find that $\tilde{\pi} \in \mathcal{A}(\tilde{x})$ and $X_T^{\tilde{x},\tilde{\pi}} = Y \geq H$D Thus $\tilde{x} \geq \Pi(H)$D

On the other hand, suppose that $X_T^{x,\pi} \geq H$ for some $\pi \in \mathcal{A}(x)$. Then, from the supermartingale property of $\{L_t^\kappa X_t^{x,\pi}\}$,

$$\mathbb{E}[L_T^\kappa H] \leq \mathbb{E}[L_T^\kappa X_T^{x,\pi}] \leq x, \quad \kappa \in \mathcal{D}. \tag{2.3}$$

It follows from $H = \delta Y + (1-\delta)Y1_{\{\tau>T\}}$ that the left-hand side in (2.3) can be written as

$$\mathbb{E}[L_T^\kappa H] = \mathbb{E}[B_T^{-1} Z_T^* Z_T^\kappa \delta Y] + \mathbb{E}[L_T^\kappa (1-\delta) Y 1_{\{\tau>T\}}]. \tag{2.4}$$

Since the quadratic covariation of $\{Z_t^\kappa\}$ and an $\mathbb{F}$-martingale is equal to zero, the process $Z_t^\kappa \mathbb{E}[B_T^{-1} Z_T^* \delta Y | \mathcal{F}_t]$ is a local martingale. So, if Y is bounded then this process is a martingale. Therefore, by approximating Y with $Y \wedge n$ and by the monotone convergence theorem, we find that the first term in the right-hand side in (2.4) is given by $\mathbb{E}[B_T^{-1} Z_T^* \delta Y]$. From this and (2.3) we have

$$\mathbb{E}[B_T^{-1} Z_T^* \delta Y] + \sup_{\kappa \in \mathcal{D}} \mathbb{E}[L_T^\kappa (1-\delta) Y 1_{\{\tau>T\}}] \leq \Pi(H).$$

However, for any constant $\kappa > -1$,

$$\begin{aligned}
\mathbb{E}[L_T^\kappa Y 1_{\{\tau>T\}}] &= \mathbb{E}[B_T^{-1} Z_T^* Y (1 + \kappa 1_{\{\tau \leq T\}}) e^{-\kappa \int_0^{\tau \wedge T} \mu_t dt} 1_{\{\tau>T\}}] \\
&= \mathbb{E}[B_T^{-1} Z_T^* Y 1_{\{\tau>T\}} e^{-\kappa \int_0^T \mu_t dt}] = \mathbb{E}[B_T^{-1} Z_T^* Y G_T e^{-\kappa \int_0^T \mu_t dt}] \\
&= \mathbb{E}[B_T^{-1} Z_T^* Y e^{-(\kappa+1) \int_0^T \mu_t dt}].
\end{aligned}$$

Hence $\mathbb{E}[L_T^\kappa Y 1_{\{\tau>T\}}] \to \mathbb{E}[B_T^{-1} Z_T^* Y]$ as $\kappa \searrow -1$. Thus the proposition follows. □

3. Maximizing the probability of the super-hedge

Proposition 2.1 implies that if the hedger of the defaultable claim wants to hedge the claim almost surely then s/he needs to have the perfect hedging cost for the liability to be paid when the default does not occur. However, the price of H should reflect the possibility of default and be smaller than

$\mathbb{E}^*[B_T^{-1}Y]$ since one can receive Y almost surely with this cost buying in the default-free market. In other words, an initial wealth for hedging of H may be smaller than $\mathbb{E}^*[B_T^{-1}Y]$. In such a case there is the possibility of the shortfall in hedging of H. One criterion for measuring this downside risk is the probability of super-hedging being successful. Our objective is thus to solve the following problem: for $x < \mathbb{E}^*[B_T^{-1}Y]$,

$$\max_{\pi\in\mathcal{A}(x)} \mathbb{P}(X_T^{x,\pi} \geq H). \tag{3.1}$$

Adapting an optimal portfolio for this problem as a hedging strategy for H is usually called a quantile hedging. To solve the quantile hedging problem (3.1), let us consider the Neyman–Pearson type problem defined by

$$\max_{\varphi\in\mathcal{R}} \mathbb{E}[\varphi] \tag{3.2}$$

where

$$\mathcal{R} = \{\varphi : 0 \leq \varphi \leq 1 \text{ a.s., } \sup_{L\in\mathcal{L}} \mathbb{E}[LH\varphi] \leq x\}.$$

As in [12], our problem is reduced to the Neyman–Pearson type problem via the following proposition.

Proposition 3.1. *Suppose that there exist $A \in \mathcal{G}_T$ and $\{\hat{\pi}_t\} \in \mathcal{A}(x)$ such that 1_A solves the Neyman–Pearson type problem (3.2) and $X_T^{x,\hat{\pi}} \geq H1_A$ a.s. Then $\hat{\pi}$ is optimal for the quantile hedging problem (3.1).*

Proof. For $\pi \in \mathcal{A}(x)$,

$$\begin{aligned}\mathbb{E}\left[L_T^\kappa 1_{\{X_T^{x,\pi}\geq H\}}H\right] &\leq \mathbb{E}\left[L_T^\kappa 1_{\{X_T^{x,\pi}\geq H\}}X_T^{x,\pi}\right]\\ &\leq \mathbb{E}[X_T^{x,\pi}L_T^\kappa] \leq x, \quad \kappa\in\mathcal{D}.\end{aligned} \tag{3.3}$$

Thus $1_{\{X_T^{x,\pi}\geq H\}} \in \mathcal{R}$. Hence

$$\max_{\pi\in\mathcal{A}(x)} \mathbb{P}(X_T^{x,\pi} \geq H) \leq \max_{\varphi\in\mathcal{R}} \mathbb{E}[\varphi]. \tag{3.4}$$

On the other hand, denoting $\hat{X} = X_T^{x,\hat{\pi}}$, we see

$$\mathbb{P}(\hat{X} \geq H) \geq \mathbb{P}(\hat{X} \geq H, A) = \mathbb{P}(\hat{X} \geq H1_A, A) = \mathbb{P}(A) = \max_{\varphi\in\mathcal{R}} \mathbb{E}[\varphi].$$

Combining this with (3.4), we have the proposition. □

For the reduced problem (3.2), we first consider the default-free case $\delta = 1$, i.e.,

$$H = Y. \tag{3.5}$$

In this case, for every $L \in \mathcal{L}$ and $\varphi \in \mathcal{R}$ we have $\mathbb{E}[LH\varphi] = \mathbb{E}^*[B_T^{-1}H \mathbb{E}[\varphi|\mathcal{F}_T]]$. Thus the problem (3.2) is equivalent to the following simple one:

$$\max_{\varphi \in \mathcal{R}_0} \mathbb{E}[\varphi] \tag{3.6}$$

where

$$\mathcal{R}_0 = \{\varphi : 0 \le \varphi \le 1 \text{ a.s., } \mathcal{F}_T\text{-measurable, } \mathbb{E}^*[B_T^{-1}H\varphi] \le x\}.$$

Define the probability measure $\mathbb{Q}^*$ on $(\Omega, \mathcal{G})$ by $d\mathbb{Q}^*/d\mathbb{P}^* = B_T^{-1}H/\mathbb{E}^*[B_T^{-1}H]$. Then the problem (3.6) is written as the simple testing problem:

$$\text{maximize} \int \varphi d\mathbb{P} \quad \text{subject to} \int \varphi d\mathbb{Q}^* \le \alpha := \frac{x}{\mathbb{E}^*[B_T^{-1}H]}.$$

By the classical Neyman–Pearson lemma (see, e.g., [14]), the [0, 1]-valued $\mathcal{F}_T$-measurable random variable

$$\varphi = 1_{\left\{\frac{d\mathbb{P}}{d\mathbb{P}^*} > z_0 \frac{d\mathbb{Q}^*}{d\mathbb{P}^*}\right\}} + \kappa 1_{\left\{\frac{d\mathbb{P}}{d\mathbb{P}^*} = z_0 \frac{d\mathbb{Q}^*}{d\mathbb{P}^*}\right\}}$$

is shown to be the solution to the above problem. Here,

$$z_0 = \inf\left\{z : \mathbb{Q}^*\left(\frac{d\mathbb{P}}{d\mathbb{P}^*} > z\frac{d\mathbb{Q}^*}{d\mathbb{P}^*}\right) \le \alpha\right\},$$

and

$$\kappa = \begin{cases} 0, & \text{if } \mathbb{Q}^*\left(\frac{d\mathbb{P}}{d\mathbb{P}^*} = z_0\frac{d\mathbb{Q}^*}{d\mathbb{P}^*}\right) = 0, \\ \dfrac{\alpha - \mathbb{Q}^*\left(\frac{d\mathbb{P}}{d\mathbb{P}^*} > z_0\frac{d\mathbb{Q}^*}{d\mathbb{P}^*}\right)}{\mathbb{Q}^*\left(\frac{d\mathbb{P}}{d\mathbb{P}^*} = z_0\frac{d\mathbb{Q}^*}{d\mathbb{P}^*}\right)}, & \text{if } \mathbb{Q}^*\left(\frac{d\mathbb{P}}{d\mathbb{P}^*} = z_0\frac{d\mathbb{Q}^*}{d\mathbb{P}^*}\right) > 0. \end{cases}$$

Define

$$y_0 = \frac{z_0}{\mathbb{E}^*[B_T^{-1}Y]}, \qquad \xi_0 = y_0 B_T^{-1} Z_T^* Y.$$

Then, summarizing the above argument and using Proposition 3.1, we obtain the following:

Theorem 3.2. *Let H be a random variable satisfying* (3.5) *and $0 < \mathbb{E}^*[B_T^{-1}H] < \infty$. Assume that $\mathbb{P}(\xi_0 = 1) = 0$. Then the perfect hedging portfolio for $H1_{\{\xi_0<1\}}$ solves the quantile hedging problem* (3.1).

Next we consider the defaultable case with zero recovery rate, i.e., the case where H is of the following form:

$$H = Y 1_{\{\tau > T\}}. \tag{3.7}$$

We adapt the convex duality approach in [8] and [19]. Observe that for $\varphi \in \mathcal{R}$, $y \geq 0$, $L \in \mathcal{L}$,

$$\mathbb{E}[\varphi] = E[\varphi(1 - yLH)] + y\mathbb{E}[LH\varphi] \leq \mathbb{E}[(1 - yLH)_+] + yx. \tag{3.8}$$

Thus the following dual problem naturally arises:

$$\inf_{y \geq 0,\ L \in \mathcal{L}} \{\mathbb{E}[(1 - yLH)_+] + yx\}.$$

For this duality approach to be successful, the key in [8] is to consider $\overline{\mathcal{L}}$ instead of $\mathcal{L}$, defined with our notations by

$$\begin{aligned}\overline{\mathcal{L}} = \{L \in L^1 : \mathbb{E}[B_T L] \leq 1,\ \mathbb{E}[LH] \\ \leq \sup\nolimits_{L' \in \mathcal{L}} \mathbb{E}[L'H],\ \mathbb{E}[LH\varphi] \leq x \quad (\varphi \in \mathcal{R})\},\end{aligned}$$

where $L^1 = L^1(\Omega, \mathcal{G}, \mathbb{P})$. Then the set $\overline{\mathcal{L}}$ includes $\mathcal{L}$ and is convex and closed under almost sure convergence. Then, the existence of a solution $(\hat{L}, \hat{y})$ to the modified dual problem

$$\inf_{y \geq 0, L \in \overline{\mathcal{L}}} \{\mathbb{E}[(1 - yLH)_+] + yx\}$$

in the class $\overline{\mathcal{L}} \times \mathbb{R}_+$ is guaranteed. In addition, there is no duality gap and for some $C : \Omega \to [0, 1]$ the random variable

$$\hat{\varphi} := 1_{\{\hat{y}\hat{L}H < 1\}} + C 1_{\{\hat{y}\hat{L}H = 1\}}$$

is a solution to the randomized version of the Neyman–Pearson type problem (3.8). We refer to [8] and [19] for details. If, moreover, $\mathbb{P}(\hat{y}\hat{L}H = 1) = 0$, then by Proposition 3.1 a super-hedging portfolio for $H 1_{\{\hat{y}\hat{L}H < 1\}}$ becomes a solution to our quantile hedging problem. However, in general, it is difficult to find an explicit solution in the abstract class $\overline{\mathcal{L}}$.

Consequently, for our quantile hedging problem, we need to choose a good class $\overline{\mathcal{L}}_1 \supset \mathcal{L}$ such that we can find an explicit dual solution $(\hat{L}, \hat{y}) \in \overline{\mathcal{L}}_1 \times \mathbb{R}_+$ and that the duality argument works (i.e., there causes no duality gap). Moreover we need to solve the super-hedging problem explicitly.

To this end, we first introduce the class $\overline{\mathcal{L}}_1$ defined by the closed hull of $\mathcal{L}$ with respect to L^1 norm. Since $\mathcal{L}$ is convex (see, e.g., [4]), so is $\overline{\mathcal{L}}_1$. Thus, $\overline{\mathcal{L}}_1$ is a closed convex set in L^1. Then the Neyman–Pearson problem (3.2) is replaced by

$$\max_{\varphi \in \mathcal{R}_1} \mathbb{E}[\varphi] \tag{3.9}$$

where

$$\mathcal{R}_1 = \{\varphi : 0 \le \varphi \le 1 \text{ a.s.}, \ \sup_{L \in \overline{\mathcal{L}}_1} \mathbb{E}[LH\varphi] \le x\}.$$

As in Proposition 3.1 we have the following reduction result. The proof is given in [21].

Proposition 3.3. *Suppose that there exist $A \in \mathcal{G}_T$ and $\hat{\pi} \in \mathcal{A}(x)$ such that 1_A solves the Neyman–Pearson type problem (3.9) and $X_T^{x,\hat{\pi}} \ge H1_A$ a.s. Then $\hat{\pi}$ is optimal for the quantile hedging problem (3.1).*

Then the dual problem is modified as

$$\inf_{y \ge 0, L \in \overline{\mathcal{L}}_1} \{\mathbb{E}[(1 - yLH)_+] + yx\}. \tag{3.10}$$

Define

$$\hat{L} = B_T^{-1} Z_T^* 1_{\{\tau > T\}} e^{\int_0^T \mu_t dt}. \tag{3.11}$$

Then, since $\kappa_t > -1$,

$$\begin{aligned}
\mathbb{E}[1 \wedge (yL_T^{\kappa} H)] &= \mathbb{E}[1 \wedge (yZ_T^* Z_T^{\kappa} Y B_T^{-1}) 1_{\{\tau > T\}}] \\
&= \mathbb{E}[1 \wedge (yZ_T^* e^{-\int_0^T \kappa_s \mu_s ds} Y B_T^{-1}) 1_{\{\tau > T\}}] \\
&\le \mathbb{E}[1 \wedge (yZ_T^* e^{\int_0^T \mu_s ds} Y B_T^{-1}) 1_{\{\tau > T\}}] = \mathbb{E}[1 \wedge (y\hat{L}H)].
\end{aligned}$$

This and the relation $\mathbb{E}(1 - yL_T^{\kappa}H)_+ = 1 - \mathbb{E}[1 \wedge (yL_T^{\kappa}H)]$ suggest that $\hat{L}$ is a candidate for an optimal dual variable. Indeed, we can show that the following result holds, which is proved in [21].

Theorem 3.4. *Suppose that H is as in* (3.7) *with $\mathbb{E}^*[B_T^{-1}Y] < \infty$. Then $\hat{L}$ defined by* (3.11) *solves*

$$\inf_{L \in \overline{\mathcal{L}}_1} \mathbb{E}[(1 - yLH)_+].$$

Moreover, there exists $\hat{y} > 0$ that minimizes

$$h(y) := \mathbb{E}[(1 - y\hat{L}H)_+] + yx$$

over $y \ge 0$. The pair $(\hat{y}, \hat{L})$ is optimal for the minimization problem (3.10).

Let $\hat{y}_1 = \hat{y} > 0$ be as in the previous theorem and consider the $\mathcal{F}_T$-measurable random variable ξ_1 defined by

$$\xi_1 = \hat{y}_1 B_T^{-1} Z_T^* e^{\int_0^T \mu_t dt} Y.$$

Then we can describe an optimal quantile hedging portfolio in the following way:

Theorem 3.5. *Suppose that H is as in* (3.7) *with $\mathbb{E}^*[B_T^{-1}Y] < \infty$ and that $\mathbb{P}(\xi_1 = 1) = 0$. Then the perfect hedging portfolio for $Y1_{\{\xi_1<1\}}$ is optimal for the quantile hedging problem* (3.1).

The proof of this theorem is similar to that of Theorem 4.6 and given in [21], so omitted.

Remark 3.6. The condition $\mathbb{P}(\xi_1 = 1) = 0$ is satisfied, e.g., in the case where the following are all satisfied: τ is independent to $\{W_t\}$, the market model is described by Black–Scholes one, and Y is given by a plain vanilla option. In more general Markovian cases, Hörmander's condition in Malliavin calculus (see, e.g., Nualart [22]) can be used to check that ξ_1 is continuously distributed.

Remark 3.7. Since the perfect hedging strategy for $Y1_{\{\xi_1<1\}}$ is an $\mathbb{F}$-predictable process, Theorem 3.5 gives an explicit solution to the quantile hedging problems with respect to $\mathbb{F}$-predictable portfolios studied in [26] in the case of zero-recovery rate.

4. Minimizing the expected loss of hedging

So far, we have focused on the quantile hedging problem. If we must say a shortcoming of this method, then we would mention that the quantile hedging cannot reflect the shortfall size. Taking this point into consideration, we study the minimization problem of the expected discounted shortfall. Our objective is thus to solve the following problem: for $x < \mathbb{E}^*[B_T^{-1}Y]$,

$$\min_{\pi\in\mathcal{A}(x)} \mathbb{E}[B_T^{-1}(H - X_T^{x,\pi})_+]. \tag{4.1}$$

Here H is as in (2.2) and satisfies $\mathbb{E}[B_T^{-1}H] < \infty$.

As in Sect. 3, our problem is reduced to the Neyman–Pearson type problem

$$\max_{\varphi\in\mathcal{R}} \mathbb{E}[B_T^{-1}H\varphi] \tag{4.2}$$

via the following proposition:

Proposition 4.1. *Suppose that there exist $\hat{\varphi} \in \mathcal{R}$ and $\hat{\pi} \in \mathcal{A}(x)$ such that $\hat{\varphi}$ solves the Neyman–Pearson type problem (4.2) and $X_T^{x,\hat{\pi}} \geq H1_A$ a.s. Then $\hat{\pi}$ is optimal for the problem (4.1).*

Proof. Let $\pi \in \mathcal{A}(x)$ and set $X = X_T^{x,\pi}$. Consider the success ratio

$$\varphi = 1_{\{X \geq H\}} + \frac{X}{H} 1_{\{X < H\}}.$$

Then $\varphi H = X \wedge H$ and so

$$\mathbb{E}[L\varphi H] \leq \mathbb{E}[LX] \leq x, \quad L \in \mathcal{L},$$

which implies $\varphi \in \mathcal{R}$. Hence for every $\pi \in \mathcal{A}(x)$

$$\mathbb{E}[B_T^{-1}(H - X_T^{x,\pi})_+] = \mathbb{E}[B_T^{-1} H(1 - \varphi)] \geq \min_{\varphi' \in \mathcal{R}} \mathbb{E}[B_T^{-1} H(1 - \varphi')].$$

On the other hand, it follows from $X_T^{x,\hat{\pi}} \geq H\hat{\varphi}$ that $(H - X_T^{x,\hat{\pi}})_+ \leq (H - H\hat{\varphi})_+ = H(1 - \hat{\varphi})$. Therefore

$$\mathbb{E}[B_T^{-1}(H - X_T^{x,\hat{\pi}})_+] \leq \mathbb{E}[B_T^{-1} H(1 - \hat{\varphi})] = \min_{\varphi' \in \mathcal{R}} \mathbb{E}[B_T^{-1} H(1 - \varphi')].$$

Thus the proposition follows. □

For the reduced problem (4.2), as in Sect. 3, we first consider the default-free case

$$H = Y. \tag{4.3}$$

In this case, the problem (4.2) is equivalent to the following simple one:

$$\max_{\varphi \in \mathcal{R}_0} \mathbb{E}[B_T^{-1} H\varphi]. \tag{4.4}$$

Define the probability measure $\mathbb{Q}$ and $\mathbb{Q}^*$ on $(\Omega, \mathcal{G})$ by $d\mathbb{Q}/d\mathbb{P} = B_T^{-1} H/\mathbb{E}[B_T^{-1} H]$ and $d\mathbb{Q}^*/d\mathbb{P}^* = B_T^{-1} H/\mathbb{E}^*[B_T^{-1} H]$, respectively. Then the problem (4.4) is written as the simple testing problem:

$$\text{maximize} \int \varphi d\mathbb{Q} \quad \text{subject to} \int \varphi d\mathbb{Q}^* \leq \alpha.$$

By the classical Neyman–Pearson lemma, the $[0, 1]$-valued $\mathcal{F}_T$-measurable random variable

$$\tilde{\varphi}_0 = 1_{\left\{\frac{d\mathbb{Q}}{d\mathbb{P}^*} > \tilde{z}_0 \frac{d\mathbb{Q}^*}{d\mathbb{P}^*}\right\}} + \tilde{\kappa}_0 1_{\left\{\frac{d\mathbb{Q}}{d\mathbb{P}^*} = \tilde{z}_0 \frac{d\mathbb{Q}^*}{d\mathbb{P}^*}\right\}}$$

is shown to be the solution to the above problem. Here,

$$\tilde{z}_0 = \inf\left\{z : \mathbb{Q}^*\left(\frac{d\mathbb{Q}}{d\mathbb{P}^*} > z\frac{d\mathbb{Q}^*}{d\mathbb{P}^*}\right) \leq \alpha\right\},$$

and

$$\tilde{\kappa}_0 = \begin{cases} 0, & \text{if } \mathbb{Q}^*\left(\dfrac{d\mathbb{Q}}{d\mathbb{P}^*} = \tilde{z}_0 \dfrac{d\mathbb{Q}^*}{d\mathbb{P}^*}\right) = 0, \\ \dfrac{\alpha - \mathbb{Q}^*\left(\frac{d\mathbb{Q}}{d\mathbb{P}^*} > \tilde{z}_0 \frac{d\mathbb{Q}^*}{d\mathbb{P}^*}\right)}{\mathbb{Q}^*\left(\frac{d\mathbb{Q}}{d\mathbb{P}^*} = \tilde{z}_0 \frac{d\mathbb{Q}^*}{d\mathbb{P}^*}\right)}, & \text{if } \mathbb{Q}^*\left(\dfrac{d\mathbb{Q}}{d\mathbb{P}^*} = \tilde{z}_0 \dfrac{d\mathbb{Q}^*}{d\mathbb{P}^*}\right) > 0. \end{cases}$$

Define

$$\tilde{y}_0 = \tilde{z}_0 \cdot \frac{\mathbb{E}[B_T^{-1} H]}{\mathbb{E}^*[B_T^{-1} Y]}, \qquad \tilde{\xi}_0 = \tilde{y}_0 Z_T^*.$$

Then, summarizing the above argument and using Proposition 4.1, we obtain the following:

Theorem 4.2. *Let H be a random variable satisfying (4.3) and $0 < \mathbb{E}^*[B_T^{-1} H] < \infty$. Assume that $\mathbb{P}(\tilde{\xi}_0 = 1) = 0$. Then the perfect hedging portfolio for $H 1_{\{\tilde{\xi}_0 < 1\}}$ solves the quantile hedging problem (4.1).*

Next we consider the defaultable case with zero recovery rate:

$$H = Y 1_{\{\tau > T\}}. \tag{4.5}$$

To this end, we first introduce the class $\overline{\mathcal{L}}_2$ defined by the closed hull of $\mathcal{L}$ with respect to $L^1(\mathbb{Q}') := L^1(\Omega, \mathcal{G}, \mathbb{Q}')$ norm, where $\mathbb{Q}'$ is defined by $d\mathbb{Q}'/d\mathbb{P} = H/\mathbb{E}[H]$. Then the Neyman–Pearson problem (4.2) is replaced by

$$\max_{\varphi \in \mathcal{R}_1} \mathbb{E}[B_T^{-1} H \varphi] \tag{4.6}$$

As in Proposition 4.1 we have the following reduction result. The proof is almost the same. To handle $L^1(\mathbb{Q}')$-closed hull, we can use an almost sure convergent subsequence on $\{H > 0\}$ in $\overline{\mathcal{L}}_2$ and Fatou's lemma. The details are left to the reader.

Proposition 4.3. *Suppose that there exist $\hat{\varphi} \in \mathcal{R}_1$ and $\hat{\pi} \in \mathcal{A}(x)$ such that $\hat{\varphi}$ solves the Neyman–Pearson type problem* (4.6) *and $X_T^{x,\hat{\pi}} \geq H\hat{\varphi}$ a.s. Then $\hat{\pi}$ is optimal for the problem* (4.1).

The associated dual problem is

$$\sup_{y \geq 0, L \in \overline{\mathcal{L}}_2} \{\mathbb{E}[H(B_T^{-1} \wedge yL)] - xy\}. \tag{4.7}$$

Then the following result shows that $\hat{L}$ becomes an optimal dual variable.

Proposition 4.4. *Suppose that H is as in (4.5) with $\mathbb{E}^*[B_T^{-1}Y] < \infty$. Then $\hat{L}$ defined by (3.11) solves*

$$\sup_{L \in \tilde{\mathcal{L}}} \mathbb{E}[H(B_T^{-1} \wedge yL)].$$

Moreover, there exists $\hat{y} > 0$ that maximizes

$$\mathbb{E}[H(B_T^{-1} \wedge y\hat{L})] - yx$$

over $y \geq 0$. The pair $(\hat{y}, \hat{L})$ is optimal for the maximization problem (4.7).

Let $\hat{y}_2 = \hat{y}$ be as in the above proposition. Consider the set

$$\mathcal{M} := \{(yHL, y) \in L^1 \times \mathbb{R} : L \in \overline{\mathcal{L}}_2, \quad y \geq 0\}.$$

Lemma 4.5. *The set $\mathcal{M}$ is closed and convex in the Banach space $L^1 \times \mathbb{R}$ with the norm $\|(L, y)\| := \mathbb{E}[|L|] + |y|$. Moreover, the functional*

$$L^1 \times \mathbb{R} \ni (K, y) \mapsto U(K, y) := yx - \mathbb{E}[HB_T^{-1} \wedge K]$$

is proper, convex and lower semi-continuous on $L^1 \times \mathbb{R}$, and satisfies

$$\inf_{(K,y) \in \mathcal{M}} U(K, y) = U(\hat{y}_2 H\hat{L}, \hat{y}_2).$$

Proof. First, the convexity of $\mathcal{M}$ easily follows from that of $\overline{\mathcal{L}}_2$.

Let $(y_n HL_n, y_n) \in \mathcal{M}$ be a sequence that converges to some (K, y) in $L^1 \times \mathbb{R}$. Then $y_n \to y$ and so

$$\mathbb{E}[H|y_n L_n - yL_n|] \leq |y_n - y|x^* \to 0 \quad (n \to \infty).$$

Thus, if $y > 0$ then denoting $L = (K/yH)1_{\{H>0\}}$ we have

$$\mathbb{E}[H|L_n - L|] \to 0 \quad (n \to \infty).$$

$L^1(\mathbb{Q}')$-closedness of $\overline{\mathcal{L}}_2$ implies $L \in \overline{\mathcal{L}}_2$.

If $y = 0$ then $y_n L_n \to yL$ in $L^1(\mathbb{Q}')$ for an arbitrary $L \in \overline{\mathcal{L}}_2$. Therefore $\mathcal{M}$ is closed.

The proofs of other assertions are simple, so omitted. □

Put $\xi_2 := \hat{y}_2 Z_T^* e^{\int_0^T \mu_t dt} Y$. Then we have the following:

Theorem 4.6. *Suppose that H is as in* (4.5) *with $\mathbb{E}^*[B_T^{-1}Y] < \infty$ and that $\mathbb{P}(\xi_2 = 1) = 0$. Then the perfect hedging portfolio for $Y1_{\{\xi_2<1\}}$ is optimal for problem* (4.1).

Proof. We follow essentially the same arguments as in [8] and [19]. Denote $L^\infty(\Omega, \mathcal{G}, \mathbb{P})$ by L^∞. Let us consider the normal cone

$$\mathcal{N}(\hat{y}_2\hat{L}H, \hat{y}_2) := \Big\{(\varphi, u) \in L^\infty \times \mathbb{R} : \mathbb{E}[\hat{y}_2\hat{L}H\varphi] + \hat{y}_2 u \geq \mathbb{E}[yLH\varphi] + yu, \quad (yLH, y) \in \mathcal{M}\Big\}$$

to the set $\mathcal{M}$ at $(\hat{y}_2\hat{L}H, \hat{y}_2)$, and the subdifferential

$$\partial U(\hat{y}_2\hat{L}H, \hat{y}_2) := \Big\{(\varphi, u) \in L^\infty \times \mathbb{R} : U(\hat{y}_2\hat{L}H, \hat{y}_2) - U(K, y) \leq \mathbb{E}[\varphi(\hat{y}_2\hat{L}H - K)] + u(\hat{y}_2 - y), \quad (K, y) \in L^1 \times \mathbb{R}\Big\}$$

at the same point. Then, by Corollary 4.6.3 in Aubin and Ekeland [1],

$$(0, 0) \in \partial U(\hat{y}_2\hat{L}H, \hat{y}_2) + \mathcal{N}(\hat{y}_2\hat{L}H, \hat{y}_2).$$

This implies that there exists $(\hat{\varphi}, \hat{u}) \in L^\infty \times \mathbb{R}$ such that $(\hat{\varphi}, \hat{u}) \in \mathcal{N}(\hat{y}_2\hat{L}H, \hat{y}_2)$ and $(-\hat{\varphi}, -\hat{u}) \in \partial U(\hat{y}_2\hat{L}H, \hat{y}_2)$. Hence we obtain

$$\mathbb{E}[H\hat{\varphi}(\hat{y}_2\hat{L} - yL)] + (\hat{y}_2 - y)\hat{u} \geq 0, \quad (L, y) \in \overline{\mathcal{L}}_2 \times \mathbb{R}_+, \tag{4.8}$$

$$(x + \hat{u})(\hat{y}_2 - y) \leq \mathbb{E}[\hat{\varphi}(K - \hat{y}_2\hat{L}H)] - \mathbb{E}[HB_T^{-1} \wedge K] + \mathbb{E}[HB_T^{-1} \wedge \hat{y}_2\hat{L}H], \quad (K, y) \in L^1 \times \mathbb{R}. \tag{4.9}$$

By letting $y \to \pm\infty$, we see that (4.9) holds only if $\hat{u} = -x$. From (4.8) with $\hat{u} = -x$, $y = \hat{y}_2 \pm \delta$ $(\delta > 0)$, and $L = \hat{L}$, we have

$$\mathbb{E}[H\hat{\varphi}\hat{L}] = x. \tag{4.10}$$

Thus, reading (4.8) with $y = \hat{y}_2$, we get

$$\mathbb{E}[H\hat{\varphi}L] \leq x, \quad L \in \overline{\mathcal{L}}_2. \tag{4.11}$$

Equation (4.9) is now written as

$$\mathbb{E}[\hat{\varphi}(K - \hat{y}_2\hat{L}H)] - \mathbb{E}[HB_T^{-1} \wedge K] + \mathbb{E}[HB_T^{-1} \wedge \hat{y}_2\hat{L}H] \geq 0, \quad K \in L^1. \tag{4.12}$$

Considering (4.12) for $K = \hat{y}_2\hat{L}H + 1_A$ with arbitrary $A \in \mathcal{G}$, we see that $0 \leq \mathbb{E}[\hat{\varphi}1_A]$. Thus $\hat{\varphi} \geq 0$ a.s. Similarly, considering (4.12) for $K = \hat{y}_2\hat{L}H - 1_A$ with arbitrary $A \in \mathcal{G}$ and using $(x + y)_+ \leq (x)_+ + (y)_+$ for $x, y \in \mathbb{R}$, we see that $0 \leq \mathbb{E}(1 - \hat{\varphi})1_A$. Thus $\hat{\varphi} \leq 1$ a.s. Combining with (4.11), we have $\hat{\varphi} \in \mathcal{R}$.

Equation (4.12) for $L = HB_T^{-1}$ implies $\mathbb{E}[\hat{\varphi}H(B_T^{-1} - \hat{y}_2\hat{L})] \geq \mathbb{E}H(B_T^{-1} - \hat{y}_2\hat{L})_+$. Thus $\hat{\varphi}(HB_T^{-1} - \hat{y}_2\hat{L}H) = (HB_T^{-1} - \hat{y}_2\hat{L}H)_+$ a.s. From this and $\hat{\varphi} \in \mathcal{R}$ we find that $\hat{\varphi} = 1$ on $\{\hat{y}_2\hat{L}H < B_T^{-1}\} \cap \{H > 0\}$ and $\hat{\varphi} = 0$ on $\{\hat{y}_2\hat{L}H > B_T^{-1}\} \cap \{H > 0\}$. Hence there must be some $[0, 1]$-valued random variables C and J such that the representation

$$\hat{\varphi} = 1_{\{\hat{y}_2\hat{L}H<B_T^{-1},H>0\}} + C1_{\{\hat{y}_2\hat{L}H=B_T^{-1},H>0\}} + J1_{\{H=0\}}. \tag{4.13}$$

holds. Now, by the assumption of the theorem, $\mathbb{P}(\hat{y}_2\hat{L}H = B_T^{-1}) = \mathbb{P}(\xi_2 = 1, \tau > T) = 0$. So we have $\mathbb{E}[B_T^{-1}H\hat{\varphi}] = \mathbb{E}[HB_T^{-1} \wedge \hat{y}_2\hat{L}H] - \hat{y}_2x$. This and (3.8) imply that $\hat{\varphi}$ is optimal for the Neyman-Pearson type problem and that there is no duality gap. Moreover, since $H\hat{\varphi} = Y1_{\{\xi_2<1\}}1_{\{\tau>T\}}$, we can apply Proposition 2.1 to deduce that a super-hedging portfolio for $H\hat{\varphi}$ is given by the perfect hedging portfolio for $Y1_{\{\xi_2<1\}}$. From this and Proposition 4.1, we arrive at the conclusion of the theorem. □

5. Case of non-zero recovery rate

Next we consider the case of $\delta > 0$. In this case, solving the dual problem is more difficult, and we leave it for a future study. Instead, we present solutions to our partial hedging problems in a restricted class of portfolio processes.

Notice that the payoff of the defaultable claim satisfies

$$H = \delta Y + (1 - \delta)Y1_{\{\tau>T\}} \geq \delta Y.$$

This implies that a seller of the claim must pay at least δY at the maturity. Since δY is $\mathcal{F}_T$-measurable, there exists a unique portfolio process $\{\pi_t^*\}_{0\leq t\leq T}$ such that

$$X_t^{x^*,\pi^*} = \mathbb{E}^*[B_T^{-1}B_t\delta Y|\mathcal{F}_t],$$

where $x^* = \mathbb{E}^*[B_T^{-1}\delta Y]$. In view of these considerations, we impose the following capital requirements on the wealth process:

$$X_t^{x,\pi} \geq \mathbb{E}^*[B_T^{-1}B_t\delta Y|\mathcal{F}_t], \quad 0 \leq t \leq T, \ \text{a.s.} \tag{5.1}$$

In particular, x must be at least x^*. We denote by $\mathcal{A}^*(x)$ all portfolio processes such that the corresponding wealth process with initial wealth x satisfies (5.1), and restrict ourselves the class of portfolio processes to $\mathcal{A}^*(x)$. Then let us consider the following quantile hedging problem

$$\max_{\pi\in\mathcal{A}^*(x)} \mathbb{P}(X_T^{x,\pi} \geq H) \tag{5.2}$$

as well as the minimization the expected discounted shortfall

$$\min_{\pi\in\mathcal{A}^*(x)} \mathbb{E}[B_T^{-1}(H - X_T^{x,\pi})_+]. \tag{5.3}$$

It follows from $H = \delta Y + (1-\delta)Y1_{\{\tau>T\}}$ that $\mathbb{P}(X_T^{x,\pi} \geq H) = \mathbb{P}(X_T^{x-x^*,\pi-\pi^*} \geq (1-\delta)Y1_{\{\tau>T\}})$. Thus the problem (5.2) is reduced to the maximization problem of $\mathbb{P}(X_T^{x',\pi'} \geq H')$ over all portfolio processes $\pi' \in \mathcal{A}(x')$. Here, $x' = x - x^* \geq 0$ and $H' = (1-\delta)Y1_{\{\tau>T\}}$. Thus,

$$\max_{\pi\in\mathcal{A}^*(x)} \mathbb{P}(X_T^{x,\pi} \geq H) = \max_{\pi'\in\mathcal{A}(x')} \mathbb{P}(X_T^{x',\pi'} \geq H').$$

Therefore we can apply Theorem 3.5 to the problem (5.2) and obtain the following:

Theorem 5.1. *Suppose that* $\mathbb{E}^*[B_T^{-1}Y] < \infty$, *and let* y_1' *and* ξ_1' *be defined by*

$$y_1' = \arg\min_{y\geq 0} \left\{\mathbb{E}(1 - y\hat{L}H')_+ + yx'\right\},$$

$$\xi_1' = y_1' B_T^{-1} Z_T^* e^{\int_0^T \mu_t dt}(1-\delta)Y.$$

Suppose moreover that $\mathbb{P}(\xi_1' = 1) = 0$. *Then the perfect hedging portfolio for* $\delta Y + (1-\delta)Y1_{\{\xi_1<1\}}$ *is optimal for the quantile hedging problem* (5.2).

Similarly, by applying Theorem 4.6 to the problem (5.3), we obtain the following:

Theorem 5.2. *Suppose that* $\mathbb{E}^*[B_T^{-1}Y] < \infty$, *and let* y_2' *and* ξ_2' *be defined by*

$$y_2' = \arg\max_{y\geq 0} \left\{\mathbb{E}[H'(B_T^{-1} \wedge y\hat{L})] - yx'\right\},$$

$$\xi_2' = y_2' Z_T^* e^{\int_0^T \mu_t dt}(1-\delta)Y.$$

Suppose moreover that $\mathbb{P}(\xi_2' = 1) = 0$. *Then the perfect hedging portfolio for* $\delta Y + (1-\delta)Y1_{\{\xi_2<1\}}$ *is optimal for the problem* (5.3).

6. Conclusion

In this paper, we have studied the partial hedging problems for defaultable claims in incomplete markets modeled by Itô processes, in the case where the portfolio processes are adapted to the full filtration. The maximization of the probability of super-hedge and the minimization of the expected discounted loss of hedging are considered. The convex duality method as in [8] and [19] has played a key role to solve the reduced Neyman–Pearson type problems.

In our future research, we wish to focus on various extensions of the results obtained in this paper. Possible future research topics include

- The non-zero recovery case.
- The case where the hedging terminal time is equal to the default time.
- Multiple default times.
- Convex (coherent) risk measures as the criterion.
- Hedging by defaultable assets (e.g., CDS).

References

1. Aubin, J., Ekeland, J.: Applied Nonlinear Analysis. Wiley, New York (1984)
2. Bellamy, N., Jeanblanc, M.: Incompleteness of markets driven by a mixed diffusion. Finance Stochast. **4**, 209–222 (2000)
3. Bielecki, T.R., Rutkowski, M.: Credit Risk: Modeling, Valuation and Hedging. Springer, Berlin (2004)
4. Brémaud, P.: Point Processes and Queues: Martingale Dynamics. Springer, New York (1981)
5. Browne, S.: Reaching goals by a deadline: digital options and continuous-time active portfolio management. Adv. Appl. Probab. **31**, 551–577 (1999)
6. Cvitanić, J.: Minimizing expected loss of hedging in incomplete and constrained markets. SIAM J. Contr. Optim. **38**, 1050–1066 (2000)
7. Cvitanić, J., Karatzas, I.: On dynamic measures of risk. Finance Stochast. **3**, 904–950 (1999)
8. Cvitanić, J., Karatzas, I.: Generalized Neyman–Pearson lemma via convex duality. Bernoulli **7**, 79–97 (2001)
9. Cvitanić, J., Pham, H., Touzi, N.: A closed-form solution for the problem of super-replication under transaction costs. Finance Stochast. **3**, 35–54 (1999)
10. Cvitanić, J., Pham, H., Touzi,N.: Super-replication in stochastic volatility models with portfolio constraints. J. Appl. Probab. **36**, 523–545 (1999)
11. Davis, M.H., Clark, J.M.C.: A note on super replicating strategies. Philos. Trans. R. Soc. Lond. Ser. A Math. Phys. Eng. Sci. **347**, 485–494 (1994)
12. Föllmer, H., Leukert, P.: Quantile hedging. Finance Stochast. **3**, 251–273 (1999)
13. Föllmer, H., Leukert, P.: Efficient hedging: cost versus shortfall risk. Finance Stochast. **4**, 117–146 (2000)
14. Föllmer, H., Schied, A.: Stochastic Finance: An Introduction in Discrete Time, 2nd edn. Walter de Gruyter, Berlin (2004)

15. Jakubenas, P., Levental, S., Ryznar, M.: The super-replication problem via probabilistic methods. Ann. Appl. Probab. **13**, 742–773 (2003)
16. Kulldorff, M.: Optimal control of a favourable game with a time-limit. SIAM J. Contr. Optim. **31**, 52–69 (1993)
17. Levental, S., Skorohod, A.V.: On the possibility of hedging options in the presence of transaction costs. Ann. Appl. Probab. **7**, 410–443 (1997)
18. Nakano, Y.: Efficient hedging with coherent risk measure. J. Math. Anal. Appl. **293**, 345–354 (2004)
19. Nakano, Y.: Minimizing coherent risk measures of shortfall in discrete-time models under cone constraints. Appl. Math. Finance **10**, 163–181 (2003)
20. Nakano, Y.: Minimization of shortfall risk in a jump-diffusion model. Stat. Probab. Lett. **67**, 87–95 (2004)
21. Nakano, Y.: Quantile hedging for defaultable claims. In: Recent Advances in Financial Engineering: Proceedings of the KIER-TMU International Workshop on Financial Engineering 2009. World Scientific, 219–230 (2010)
22. Nualart, D.: The Malliavin Calculus and Related Topics, 2nd edn. Springer, Berlin (2006)
23. Pham, H.: Dynamic L^p-hedging in discrete time under cone constraints. SIAM J. Contr. Optim. **38**, 665–682 (2000)
24. Rudloff, B.: Convex hedging in incomplete markets. Appl. Math. Finance **14**, 437–452 (2007)
25. Rudloff, B.: Coherent hedging in incomplete markets. Quant. Finance **9**, 197–206 (2009)
26. Sekine, J.: Quantile hedging for defaultable securities in an incomplete market, Mathematical Economics (Kyoto, 1999), Sūrikaisekikenkyūsho Kōkyūroku, No. 1165, pp. 215–231 (2000)
27. Sekine, J.: Dynamic minimization of worst conditional expectation of shortfall. Math. Finance **14**, 605–618 (2004)
28. Soner, H.M., Shreve, S.E., Cvitanić, J.: There is no nontrivial hedging portfolio for option pricing with transaction costs. Ann. Appl. Probab. **5**, 327–355 (1995)
29. Spivak, G., Cvitanić, J.: Maximizing the probability of a perfect hedging. Ann. Appl. Probab. **9**, 1303–1328 (1999)

Adv. Math. Econ. 14, 147–181 (2011)

Advances in
MATHEMATICAL
ECONOMICS

Robust utility maximization with unbounded random endowment

Keita Owari*

Institute of Economic Research, Hitotsubashi University, 2-1 Naka, Kunitachi, Tokyo 186-8603, Japan
(e-mail: keita.owari@gmail.com)

Received: February 22, 2010
Revised: July 27, 2010

JEL classification: C61, G11

Mathematics Subject Classification (2010): 91G80, 60H30, 46N10, 60G44

Abstract. This paper studies the problem of robust utility maximization with random endowment. When the endowment is possibly unbounded, but satisfies certain integrability conditions, we first prove the fundamental duality relation between the utility maximization and the dual problem, and the existence of a solution to the dual problem. Then the existence of an optimal strategy in a certain choice of admissible class is discussed. As an application, we introduce a robust version of utility indifference prices.

Key words: Robust Utility Maximization, Convex Duality Method, Utility Indifference Valuation

1. Introduction

This paper addresses the convex duality theory for robust utility maximization, in the presence of *random endowment*. Suppose we are given a semimartingale S describing the evolution of prices of tradable assets, a class Θ

* The author deeply thanks an anonymous referee for a number of valuable suggestions. Also, the financial support from the Global Center of Excellence (COE) program "the Research Unit for Statistical and Empirical Analysis in Social Sciences (G-COE Hi-Stat)" of Hitotsubashi University is greatly acknowledged.

S. Kusuoka, T. Maruyama (eds.), *Advances in Mathematical Economics Volume 14*,
DOI: 10.1007/978-4-431-53883-7_7,

of admissible integrands for S, a utility function U, a set $\mathcal{P}$ of probability measures, and a random variable B. Then the problem is to:

$$\text{maximize} \quad \inf_{P \in \mathcal{P}} E^P[U(\theta \cdot S_T + B)], \quad \text{among } \theta \in \Theta. \qquad (1.1)$$

The set $\mathcal{P}$ is a mathematical expression of *model uncertainty* (also called *Knightian uncertainty*, in Economics). Each element $P \in \mathcal{P}$ is considered a candidate model of financial market, and the case where $\mathcal{P}$ is a singleton corresponds to the classical situation without uncertainty. See Föllmer et al. [13] for more background information about model uncertainty and robust utility maximization.

In the present paper, we consider (1.1) in the case where the utility function U is defined on the whole real line, and the endowment B is possibly unbounded but satisfies a certain uniform integrability condition in the split of [28]. We will give (1) the duality between the problem (1.1) and a minimization problem of a *robust divergence functional* over a class of local martingale measures for S (the dual problem), (2) the existence and characterizations of a solution to the dual problem, and (3) the existence of optimal strategy for (1.1) with a certain choice of Θ.

One of our primal motivation for studying the problem (1.1) is to give a way of pricing contingent claims in incomplete financial market with model uncertainty. To this aim, we will define and compute the *robust utility indifference price* of a claim B. This is mathematically a direct corollary of our duality result presented below, but possesses some reasonable properties as fair price, even under model uncertainty.

We close the introduction with a brief literature review. When $\mathcal{P}$ is a singleton (i.e., no model uncertainty), the convex duality theory for utility maximization is studied by many authors, e.g., the case of no endowment ($B \equiv 0$) by Kramkov and Schachermayer [22, 23] and Schachermayer [32, 33], the case of bounded claim by Bellini and Frittelli [4], the case of exponential utility with integrability conditions on B by Delbaen et al. [10], Kabanov and Stricker [21] and Becherer [3]. Also, Owen and Žitković [29], Biagini et al. [6] and Owari [27] deal with the case of general utility functions defined on $\mathbb{R}$ with unbounded endowment.

There is also a vast literature on robust utility maximization without endowment. When the utility function is defined on $\mathbb{R}_+$, Quenez [30] considered the problem under rather stronger assumptions including the equivalence of all $P \in \mathcal{P}$, examining the case of Brownian filtration with logarithmic and power utilities by means of BSDE. More general duality theory is developed by Schied and Wu [35] and Schied [34], followed by Hernández-Hernández and Schied [16, 17] with explicit examples in the setting of a stochastic factor model. For recent developments in this direction, see Föllmer et al. [13] and references there-in. Also, Föllmer and Gundel [12] gives a general existence result for the so called *robust f-projection* problem, which is the dual of robust utility maximization.

For the problem of the type (2.14), Owari [28] extends the duality theory of [10] to the case with model uncertainty (see also [26] for a solvable example), and this paper is an extension of [28] to a slightly general class of utility functions. Müller [25] also studies the robust exponential utility maximization with *bounded* endowment in a Brownian setting. Müller [25] solves the problem with a direct BSDE argument without duality. More recently, Wittmüss [37] studies the robust utility maximization from consumption with bounded random endowment for general utility functions defined on the positive half line.

Some more detailed information will be given after the proofs.

2. Main results

2.1. Setup

Suppose we are given a complete probability space $(\Omega, \mathcal{F}, \mathbb{P})$ equipped with a filtration $\mathbb{F} = (\mathcal{F}_t)_{t\in[0,T]}$ satisfying the usual conditions of right-continuity and $\mathbb{P}$-completeness, where $T \in (0, \infty)$ is a fixed time horizon. We assume $\mathcal{F} = \mathcal{F}_T$ for simplicity of notation. The expectation of a random variable X under $\mathbb{P}$ is denoted simply by $E[X]$, while we explicitly write $E^P[X]$ for expectation under other probability P.

Let S be a d-dimensional càdlàg *locally bounded* semimartingale on $(\Omega, \mathcal{F}, \mathbb{F}, \mathbb{P})$. For the set of admissible strategies, we will *basically* consider:

$$\Theta_{bb} := \{\theta \in L(S) : \theta_0 = 0,\ \theta \cdot S \text{ is uniformly bounded from below}\}, \quad (2.1)$$

where $L(S) := L(S, \mathbb{P})$ denotes the set of all d-dimensional predictable $(S, \mathbb{P})$-integrable processes. For the precise definitions and basic properties of stochastic integral and the space $L(S, \mathbb{P})$ (or more generally, $L(S, P)$ with $P \ll \mathbb{P}$), we refer the reader to Jacod [20] and [19, Ch. II, IV].

Utility functions and conjugate

In this paper we consider utility functions U defined on the *whole real line*, i.e., $U(x)$ is finite for all $x \in \mathbb{R}$. More precisely, we assume:

(A1) $U : \mathbb{R} \to \mathbb{R}$ is continuously differentiable, increasing, and strictly concave function satisfying the *Inada condition*:

$$\lim_{x\to-\infty} U'(x) = +\infty, \quad \text{and} \quad \lim_{x\to+\infty} U'(x) = 0. \quad (2.2)$$

(A2) U satisfies the condition of *reasonable asymptotic elasticity*:

$$AE_{-\infty}(U) := \liminf_{x\searrow-\infty} \frac{xU'(x)}{U(x)} > 1, \ AE_{+\infty}(U) := \limsup_{x\nearrow+\infty} \frac{xU'(x)}{U(x)} < 1. \tag{2.3}$$

We often need a stronger assumption than (A1):

(A1′) U satisfies (A1) and $U(+\infty) := \sup_{x\in\mathbb{R}} U(x) < \infty$.

A typical example of utility function satisfying (A1′) and (A2) is the exponential utility: $U(x) = -e^{-\alpha x}$, where $\alpha > 0$ is a risk aversion parameter.

For a given utility function U, its conjugate V is defined by

$$V(y) := \sup_{x\in\mathbb{R}}(U(x) - xy), \quad y \in \mathbb{R}. \tag{2.4}$$

In the standard language of convex analysis, V is the convex conjugate (also called the *Fenchel–Legendre transform*) of the convex function $\Phi(x) = -U(-x)$.

By (A1), V is also differentiable on $(0, \infty)$ with $V'(y) = -(U')^{-1}(y)$, and

$$V'(0) := \lim_{y\downarrow 0} V'(y) = -\infty \quad V'(+\infty) := \lim_{y\to+\infty} V'(y) = +\infty,$$

by the Inada condition, hence V is bounded from below. Also, $V(0) = U(\infty)$.

The assumption (A2) of reasonable asymptotic elasticity is equivalent to any of the following two conditions (Frittelli and Rosazza Gianin [14], Proposition 1):

1. For any compact interval $[a, b] \subset (0, \infty)$, there exist $C_1, C_2 > 0$ such that

$$V(\lambda y) \leq C_1 V(y) + C_2(y + 1), \quad \forall\lambda \in [a, b], \ y > 0. \tag{2.5}$$

2. There exist $C_1', C_2' > 0$ such that

$$y|V'(y)| \leq C_1' V(y) + C_2'(y + 1), \quad \forall y > 0. \tag{2.6}$$

Measures and V-divergences

In this paper, we always identify any set $\mathcal{Q}$ of positive finite measures absolutely continuous w.r.t. $\mathbb{P}$ with a subset of $L^1 = L^1(\Omega, \mathcal{F}, \mathbb{P})$ by identifying $\nu \in \mathcal{Q}$ with $d\nu/d\mathbb{P} \in L^1$. Also, when we speak of *weak topology* (e.g. weakly closed, weakly compact), we mean $\sigma(L^1, L^\infty)$-topology on L^1.

For any pair of positive finite measures (ν, μ), we denote by $d\nu/d\mu$ the Radon–Nikodym density of ν w.r.t. μ in the sense of Lebesgue decomposition:

$$\frac{d\nu}{d\mu} := \frac{d\nu/d\mathbb{P}}{d\mu/d\mathbb{P}} 1_{\{d\mu/d\mathbb{P}>0\}} + \infty 1_{\{d\mu/d\mathbb{P}=0, d\nu/d\mathbb{P}>0\}}.$$

To the conjugate V defined by (2.4), we associate another map on $\mathbb{R}^2_+$ by

$$V(x, y) := \begin{cases} 0 & \text{if } x = y = 0, \\ x \lim_{z\to+\infty} \frac{V(z)}{z} & \text{if } x > 0,\ y = 0, \\ yV(x/y) & \text{if } y > 0. \end{cases} \tag{2.7}$$

The map $V(\cdot, \cdot)$ is lower semicontinuous on $\mathbb{R}^2_+$. Then the V-divergence is defined for any pair of positive measures (ν, μ) by

$$V(\nu|\mu) := E[V(d\nu/d\mathbb{P}, d\mu/d\mathbb{P})]. \tag{2.8}$$

$(\nu, \mu) \mapsto V(\nu|\mu)$ is convex and weakly lower semicontinuous on $L^1_+ \times L^1_+$ (see [12, Lemma 2.7]).

Let $\mathcal{P}$ be a set of probability measures absolutely continuous w.r.t. $\mathbb{P}$, expressing the *model uncertainty*. For this set, we assume:

(A3) $\mathcal{P}$ is convex and weakly compact as a subset of L^1.

Given such a $\mathcal{P}$, we define the *robust V-divergence* functional by $\nu \mapsto V(\nu|\mathcal{P}) := \inf_{P\in\mathcal{P}} V(\nu|P)$. The functional $\nu \mapsto V(\nu|\mathcal{P})$ is also convex by the convexity of $\mathcal{P}$.

Here we recall a characterization of weak compactness in L^1 for *convex* sets, which is a combination of the Dunford–Pettis theorem and the fact that the $\sigma(L^1, L^\infty)$-topology shares the same closed *convex* sets with the strong (norm) topology.

Lemma 2.1. *A convex set $A \subset L^1$ is weakly compact if and only if it is norm closed and uniformly integrable.*

Proof. By definition, A is weakly compact if and only if it is closed and relatively compact for the $\sigma(L^1, L^\infty)$-topology. The $\sigma(L^1, L^\infty)$-relative compactness coincides with the uniform integrability by the Dunford–Pettis theorem [11, Theorem II.25], while A is weakly closed if and only if strongly closed since it is convex (see e.g. [1, Theorem 5.98]). □

A probability measure Q is called a P-absolutely continuous (resp. P-equivalent) local martingale measure for S if $Q \ll P$ (resp. $Q \sim P$) and S is a Q-local martingale. The set of P-absolutely continuous (resp. equivalent) local martingale measures is denoted by $\mathcal{M}_{loc}(P)$ (resp. $\mathcal{M}^e_{loc}(P)$), and we write $\mathcal{M}_{loc} = \mathcal{M}_{loc}(\mathbb{P})$ (resp. $\mathcal{M}^e_{loc} = \mathcal{M}^e_{loc}(\mathbb{P})$). Define also:

$$\mathcal{M}_V := \{Q \in \mathcal{M}_{loc} : V(Q|\mathcal{P}) < \infty\}, \tag{2.9}$$

and set $\mathcal{M}^e_V := \{Q \in \mathcal{M}_V : Q \sim \mathbb{P}\}$. Note that $\mathcal{M}_{loc}$ is convex and closed as a subset of $L^1(\mathbb{P})$ since S is locally bounded (see e.g., Delbaen and Schachermayer [8], Theorem 5.4). Also, $\mathcal{M}^e_{loc}$, $\mathcal{M}_V$ and $\mathcal{M}^e_V$ are convex since $V(\cdot|\mathcal{P})$ is convex.

We assume a kind of no-arbitrage condition:

(A4) $\mathcal{M}^e_V \neq \emptyset$.

Remark 2.2. By the condition (2.5) ($\Leftrightarrow$ (A2)), we have, for any pair (Q, P) of probability measures, that $V(Q|P) < \infty$ if and only if $V(\lambda Q|P) < \infty$ for all $\lambda > 0$. Therefore, $Q \in \mathcal{M}_{loc}$ is in $\mathcal{M}_V$ if and only if $V(\lambda Q|\mathcal{P}) < \infty$ for all $\lambda > 0$.

(A4) implies the existence of a pair $(Q_0, P_0) \in \mathcal{M}_V \times \mathcal{P}$ such that $Q_0 \sim P_0 \sim \mathbb{P}$, and $V(Q_0|P_0) < \infty$. In particular, $\mathcal{P} \sim \mathbb{P}$ in the sense that for all $A \in \mathcal{F}$,

$$P(A) = 0,\ \forall P \in \mathcal{P} \quad \Leftrightarrow \quad \mathbb{P}(A) = 0$$

This yields a kind of no-arbitrage condition of the following type: for any $\theta \in \Theta_{bb}$, if $P(\theta \cdot S_T \geq 0) = 1$ for all $P \in \mathcal{P}$, then $P(\theta \cdot S_T = 0) = 1$ for all $P \in \mathcal{P}$. Indeed, $P(\theta \cdot S_T \geq 0) = 1$ for all $P \in \mathcal{P} \Leftrightarrow \mathbb{P}(\theta \cdot S_T \geq 0) = 1 \Leftrightarrow Q_0(\theta \cdot S_T \geq 0) = 1 \Leftrightarrow Q_0(\theta \cdot S_T = 0) = 1$ (since S is a Q_0-local martingale and $\theta \cdot S_T$ is bounded from below) $\Leftrightarrow P(\theta \cdot S_T = 0) = 0$ for all $P \in \mathcal{P}$. However, arbitrage in the usual sense is possible under *each model* $P \in \mathcal{P}$.

Random endowment

For the random endowment B, we assume:

(A5) B is $\mathcal{F}_T$-measurable and there exists $\varepsilon > 0$ such that

$$\inf_{P \in \mathcal{P}} E^P[U(-\varepsilon B^+)] > -\infty, \tag{2.10}$$

$$\{U(-(1+\varepsilon)B^-)(dP/d\mathbb{P})\}_{P \in \mathcal{P}} \text{ is uniformly integrable.} \tag{2.11}$$

For a positive finite measure ν and a random variable X, we denote $\nu(X) := \int_\Omega X d\nu$, whenever the integral makes sense. In the dual problem, we will work with the functional $\nu \mapsto V(\nu|\mathcal{P}) + \nu(B)$. As B is allowed to be unbounded, we have to check that this functional is well-defined.

By the definition of V, we have $YD \le V(Y) - U(-D)$ for any positive random variables Y and D. In particular, taking $Y = d\nu/dP$,

$$\nu(D) \le V(\nu|P) - E^P[U(-D)] \le V(\nu|P) - \inf_{P' \in \mathcal{P}} E^{P'}[U(-D)]. \quad (2.12)$$

Taking $D = \varepsilon B^+$ and $D = (1+\varepsilon)B^-$, assumption (A5) implies that $B \in L^1(\nu)$ whenever $V(\nu|\mathcal{P}) < \infty$. Noting that $\mathrm{cone}(\mathcal{M}_V) := \{\lambda Q : \lambda \ge 0,\ Q \in \mathcal{M}_V\}$ is written as

$$\mathrm{cone}(\mathcal{M}_V) = \{0\} \cup \{\nu \in \mathrm{cone}\mathcal{M}_{loc} : V(\nu|\mathcal{P}) < \infty\}, \quad (2.13)$$

by Remark 2.2, the functional $\nu \mapsto V(\nu|\mathcal{P}) + \nu(B)$ is well-defined on $\mathrm{cone}(\mathcal{M}_V)$ by (A2) and (A5) (the case $\nu = 0$ is trivial since then $V(\nu|\mathcal{P}) + \nu(B) = V(0)$).

Remark 2.3. Assumptions (A3) and (A5) seem to depend on the choice of *reference measure* $\mathbb{P}$, but actually not: these assumptions remain true even if $\mathbb{P}$ is replaced by another probability $\mathbb{P}' \sim \mathbb{P}$. For example, the family $\mathcal{P}$ is uniformly integrable if and only if for any $\varepsilon > 0$, there exists a $\delta > 0$ such that $\mathbb{P}(A) < \delta$ implies $\sup_{P \in \mathcal{P}} P(A) \le \varepsilon$. Since $\{d\mathbb{P}/d\mathbb{P}'\}$ is clearly $\mathbb{P}'$-uniformly integrable, $\{dP/d\mathbb{P}' : P \in \mathcal{P}\}$ is again $\mathbb{P}'$-uniformly integrable.

Important Note on Assumptions. In the sequel, (A1)–(A5) are always in force (excepting Sect. 2.3) as the *standing assumptions*, and we do not cite them in the statements. On the other hand, we explicitly cite (A1′) when it is used.

2.2. Main theorems

Recall that the primal problem of this paper is:

$$\text{maximize} \quad \inf_{P \in \mathcal{P}} E^P[U(\theta \cdot S_T + B)], \quad \text{over all } \theta \in \Theta_{bb}. \quad (2.14)$$

The dual of the problem (2.14) is now stated as:

$$\text{minimize} \quad V(\lambda Q|P) + \lambda E^Q[B], \quad \text{over } \lambda > 0,\ (Q, P) \in \mathcal{M}_V \times \mathcal{P}. \quad (2.15)$$

This problem has several equivalent forms. Note that the assumption (A3) of weak compactness together with the lower semicontinuity of the V-divergence imply that the infimum $V(\nu|\mathcal{P}) = \inf_{P \in \mathcal{P}} V(\nu|P)$ is always attained by some $P_\nu \in \mathcal{P}$. Therefore, the problem (2.15) is equivalent to the minimization of $(\lambda, Q) \mapsto V(\lambda Q|\mathcal{P}) + \lambda E^Q[B]$ over $(0, \infty) \times \mathcal{M}_V$, or equivalently,

$$\text{minimize} \quad V(\nu|\mathcal{P}) + \nu(B), \quad \text{over } \nu \in \mathrm{cone}(\mathcal{M}_V) \setminus \{0\}. \quad (2.16)$$

The formulation (2.15) may be familiar and conformable with other existing research, while the second formulation (2.16) is convenient for discussing abstract results. Thus we will make proper use of two equivalent formulations depending on the situation.

Here we state all the main results of the paper, which will be proved in subsequent sections. Recall that (A1)–(A5) are always assumed without mentioning. The first one concerns the relation between the utility maximization and the dual problem:

Theorem 2.4 (Duality). *Assume (A1′). Then the following duality equality holds:*

$$\begin{aligned}\sup_{\theta\in\Theta_{bb}}\inf_{P\in\mathcal{P}} E^P[U(\theta\cdot S_T+B)] &= \inf_{\lambda>0}\inf_{(Q,P)\in\mathcal{M}_V\times\mathcal{P}}(V(\lambda Q|P)+\lambda E^Q[B])\\ &= \inf_{\nu\in\mathrm{cone}(\mathcal{M}_V)\setminus\{0\}}(V(\nu|\mathcal{P})+\nu(B)).\end{aligned} \tag{2.17}$$

Using this duality, we can compute the maximal admissible utility via the dual problem. This will ease the computation of the robust utility indifference prices as in the subjective case (see Sect. 2.3). Also, (2.17) justifies the term *dual problem.*

Next we consider the dual problem (2.16).

Theorem 2.5 (Existence for the Dual Problem). *The problem (2.16) admits a solution* $\hat{\nu}\in\mathrm{cone}(\mathcal{M}_V)\setminus\{0\}$, *i.e.,*

$$V(\hat{\nu}|\mathcal{P})+\hat{\nu}(B)=\inf_{\nu\in\mathrm{cone}(\mathcal{M}_V)\setminus\{0\}}(V(\nu|\mathcal{P})+\nu(B)). \tag{2.18}$$

Moreover, there exists a solution having the maximal support among all solutions.

As noted above, the existence of a solution $\hat{\nu}=\hat{\lambda}\hat{Q}$ already gives a minimizer $(\hat{\nu},\hat{P})=(\hat{\lambda}\hat{Q},\hat{P})$ of $(\nu,P)\mapsto V(\nu|P)+\nu(B)$ on $(\mathrm{cone}(\mathcal{M}_V)\setminus\{0\})\times\mathcal{P}$. Then the pair $(\hat{\nu},\hat{P})$ is also called a solution to the dual problem.

When $\mathcal{P}$ is a singleton (say $\{\mathbb{P}\}$), we have further that $\hat{\nu}=\hat{\lambda}\hat{Q}$ is unique, and is equivalent to $\mathbb{P}$. In the robust case, however, we can hope neither the uniqueness nor the equivalence any more, and the best we can do is to construct a solution with the maximal support, which has a kind of uniqueness in a weaker sense: the density $d\hat{\nu}/d\hat{P}$ is a.s. unique. See Proposition 4.7 below.

Aside from the generality, it sometimes happens in special cases that we can take a common pair $(\hat{Q},\hat{P})\in\mathcal{M}_V\times\mathcal{P}$ which minimizes $(Q,P)\mapsto V(\lambda Q|P)+\lambda E^Q[B]$ for all $\lambda>0$. In such a case, the function $v(\lambda)=\inf_{(Q,P)\in\mathcal{M}_V\times\mathcal{P}}(V(\lambda Q|P)+\lambda E^Q[B])$ is strictly convex, hence the *scaling factor* $\hat{\lambda}$ of $\hat{\nu}=\hat{\lambda}\hat{Q}$ is uniquely determined.

Example 2.6 (Exponential Utility). Let $U(x) = -e^{-x}$. This is called the exponential utility with the risk aversion parameter 1. Then $V(y) = y\log y - y$, hence

$$V(\lambda Q|P) + \lambda E^{Q}[B] = \lambda(\mathcal{H}(Q|P) + E^{Q}[B]) + \lambda\log\lambda - \lambda,$$

where $\mathcal{H}(Q|P)$ is the relative entropy of Q w.r.t. P. Therefore, the dual problem amounts to minimize $\mathcal{H}(Q|P) + E^{Q}[B]$ in (Q, P), and the unique minimizer $\hat{\lambda}$ is given by $\hat{\lambda} = \exp(-\inf_{(Q,P)\in\mathcal{M}_V\times\mathcal{P}}(\mathcal{H}(Q|P) + E^{Q}[B]))$. See Owari [28] for detail.

Once the existence of a solution is obtained, our next interest is in the characterization of solutions.

Theorem 2.7 (Characterization for Dual Optimizers). *Let* $(\hat{\nu}, \hat{P}) = (\hat{\lambda}\hat{Q}, \hat{P})$ *be a solution to the dual problem. Then:*

(a) $\hat{Q} \sim \hat{P}$.
(b) There exists an $(S, \hat{Q})$*-integrable predictable process* $\hat{\theta}$ *with* $\theta_0 = 0$ *such that* $\hat{\theta}\cdot S$ *is a* $\hat{Q}$*-martingale, and*

$$-V'\left(\hat{\lambda}\frac{d\hat{Q}}{d\hat{P}}\right) = \hat{\theta}\cdot S_T + B, \quad \hat{Q}\text{-a.s.} \tag{2.19}$$

We now go on to the discussion of optimal strategy. Unfortunately, the class Θ_{bb} is too small to admit an optimizer, even if $\mathcal{P} = \{\mathbb{P}\}$ and $B \equiv 0$ (see e.g., [33]). Thus, we need to enlarge the admissible class.

We proceed as follows. We first show that the duality (2.17) is *stable enough* under the change of admissible class Θ. This implies in particular that the maximal admissible utility is unchanged under certain changes of Θ. Then the following special class will turn out to admit an optimal strategy, under an additional assumption:

$$\Theta_{V,\hat{P}} := \Big\{\theta \in L(S) : \theta_0 = 0,\ \theta\cdot S \text{ is a } Q\text{-supermartingale}, \\ \forall Q \in \mathcal{M}_V(\hat{P})\Big\}, \tag{2.20}$$

where $\mathcal{M}_V(\hat{P}) = \{Q \in \mathcal{M}_V : V(Q|\hat{P}) < \infty\}$. Of course, $\Theta_{bb} \subset \Theta_{V,\hat{P}}$ by Ansel and Stricker [2], Corollary 3.5. Note also that if $\theta\cdot S$ is well-defined under $\mathbb{P}$, then it is also well-defined under any $P \ll \mathbb{P}$.

Theorem 2.8 (Optimal Strategy). *Assume (A1′).*

(a) Let $\Theta \subset L(S)$ *be such that* $\Theta_{bb} \subset \Theta \subset \Theta_{V,\hat{P}}$*. Then*

$$\begin{aligned}\sup_{\theta\in\Theta}\inf_{P\in\mathcal{P}} E^{P}[U(\theta\cdot S_T + B)] &= \sup_{\theta\in\Theta_{bb}}\inf_{P\in\mathcal{P}} E^{P}[U(\theta\cdot S_T + B)] \\ &= \inf_{\lambda>0}\inf_{(Q,P)\in\mathcal{M}_V\times\mathcal{P}}(V(\lambda Q|P) + \lambda E^{Q}[B]).\end{aligned} \tag{2.21}$$

In particular, the maximal utility is unchanged if Θ_{bb} *is replaced by* $\Theta_{V,\hat{P}}$.

(b) $\hat{\theta} \cdot S$ *is a supermartingale under all* $Q \in \mathcal{M}_V(\hat{P})$, *where* $\hat{\theta}$ *is the integrand appearing in (2.19).*

(c) *If we assume in addition that* $\hat{Q} \sim \mathbb{P}$, *then* $\hat{\theta} \in \Theta_{V,\hat{P}}$, *and the pair* $(\hat{\theta}, \hat{P}) \in \Theta_{V,\hat{P}} \times \mathcal{P}$ *is a saddle point of the map* $\Theta_{V,\hat{P}} \times \mathcal{P} \ni (\theta, P) \mapsto E^P[U(\theta \cdot S_T + B)]$, *i.e.,*

$$\min_{P\in\mathcal{P}} E^P[U(\hat{\theta}\cdot S_T+B)] = E^{\hat{P}}[U(\hat{\theta}\cdot S_T+B)] = \max_{\theta\in\Theta_{V,\hat{P}}} E^{\hat{P}}[U(\theta\cdot S_T+B)]. \tag{2.22}$$

In particular,

$$\inf_{P\in\mathcal{P}} E^P[U(\hat{\theta} \cdot S_T + B)] = \max_{\theta\in\Theta_{V,\hat{P}}} \inf_{P\in\mathcal{P}} E^P[U(\theta \cdot S_T + B)]. \tag{2.23}$$

In the assertion (c) on the saddle point argument, the "$\hat{\theta}$-part" of the saddle point is unique, while $\hat{P}$ is not, in the sense that if $(\tilde{\theta}, \tilde{P}) \in \Theta_{V,\tilde{P}} \times \mathcal{P}$ is another saddle point with $\tilde{P} \sim \mathbb{P}$, then $\hat{\theta}\cdot S = \tilde{\theta}\cdot S$ up to $\mathbb{P}$-indistinguishability. This is a consequence of the uniqueness of the density of maximal solution in the sense mentioned above.

Remark 2.9. The reader may be curious why we choose Θ_{bb} for the primal domain *even though* it can not admit an optimizer. This is a matter of universality. In the robust case, the class $\Theta_{V,\hat{P}}$ admitting an optimal strategy (if $\hat{Q} \sim \mathbb{P}$) depends on $\hat{P}$, implying that it is not available until we solve the dual problem. Furthermore, the dependence on $\hat{P}$ implies the dependence on B, which is quite undesirable when we consider the application to indifference valuation. On the other hand, Θ_{bb} is *a priori* well-defined, and depends neither on $\mathcal{P}$ nor on B. From this point of view, Theorem 2.8 (c) is understood that an optimal strategy is obtained in a "*completion*" of Θ_{bb} in a suitable sense.

2.3. Indifference valuation

We now apply our duality result to a valuation problem. Consider an investor thinking whether to buy a contingent claim B. Suppose that his/her preference is represented by the robust utility functional $X \mapsto \inf_{P\in\mathcal{P}} E^P[U(X)]$. If he/she buys the claim B at the price p, the maximal admissible utility is:

$$\sup_{\theta\in\Theta_{bb}} \inf_{P\in\mathcal{P}} E^P[U(\theta \cdot S_T - p + B)], \tag{2.24}$$

while if he/she does not buy,

$$\sup_{\theta\in\Theta_{bb}} \inf_{P\in\mathcal{P}} E^P[U(\theta \cdot S_T)]. \tag{2.25}$$

From economic point of view, the following decision seems reasonable: if (2.24) $>$ (resp. $<$) (2.25), buy (resp. not buy) the claim. If both quantities are equal, buying the claim does not change the maximal utility, hence *indifference*. A similar reasoning apply also to the *seller's* decision. Noting that $p \mapsto \sup_{\theta\in\Theta_{bb}} \inf_{P\in\mathcal{P}} E^P[U(\theta \cdot S_T - p + B)]$ is decreasing, we now arrive at the following definition.

Definition 2.10 (Indifference Prices). Buyer's *robust utility indifference price of B is defined with the convention* $\sup\emptyset = -\infty$ *by*

$$
\begin{aligned}
p_b(B) := \sup\Bigg\{ p : \ & \sup_{\theta\in\Theta_{bb}} \inf_{P\in\mathcal{P}} E^P[U(\theta \cdot S_T - p + B)] \\
& \geq \sup_{\theta\in\Theta_{bb}} \inf_{P\in\mathcal{P}} E^P[U(\theta \cdot S_T)] \Bigg\}.
\end{aligned}
$$

Also, seller's *indifference price of B is defined with the convention* $\inf\emptyset = +\infty$ *by*

$$
\begin{aligned}
p_s(B) := \inf\Bigg\{ p : \ & \sup_{\theta\in\Theta_{bb}} \inf_{P\in\mathcal{P}} E^P[U(\theta \cdot S_T + p - B)] \\
& \geq \sup_{\theta\in\Theta_{bb}} \inf_{P\in\mathcal{P}} E^P[U(\theta \cdot S_T)] \Bigg\}.
\end{aligned}
$$

In this subsection, the claim B acts as a "variable" rather than a part of "setting". Thus we drop (A5) from the "standing assumptions", and instead, we introduce:

$$
\mathcal{B} := \{B \in L^0(\Omega, \mathcal{F}_T, \mathbb{P}) : \ B \text{ satisfies (A5)}\}.
$$

This is the set of *admissible claims* for the buyer. Similarly, the set of admissible claims for the seller is $-\mathcal{B} = \{B : -B \in \mathcal{B}\}$. By (A1)–(A4), which are still in force, $\mathcal{B}$ is a convex set satisfying:

1. $L^\infty \subset \mathcal{B} \cap (-\mathcal{B})$, hence $\mathbb{R} \subset \mathcal{B} \cap (-\mathcal{B})$ in particular.
2. $B \in \mathcal{B} \Rightarrow x + B \in \mathcal{B}$ for all $x \in \mathbb{R}$. To see this, observe that

$$
U\left(-\frac{\varepsilon}{2}(x+B)^+\right) \geq \frac{1}{2}U(-\varepsilon x^+) + \frac{1}{2}U(-\varepsilon B^+).
$$

Thus, if B satisfies (2.10) with $\varepsilon > 0$, so does $x + B$ with $\varepsilon/2 > 0$. Similarly, for every $\gamma > 1$,

$$
U(-\sqrt{\gamma}(x+B)^-) \geq \frac{\sqrt{\gamma}-1}{\sqrt{\gamma}} U\left(-\frac{\gamma x^-}{\sqrt{\gamma}-1}\right) + \frac{1}{\sqrt{\gamma}} U(-\gamma B^-).
$$

Hence if B satisfies (2.11) with $\varepsilon > 0$, so does $x + B$ with $\varepsilon' = \sqrt{1+\varepsilon} - 1$.

Theorem 2.4 gives us the following *risk measure* representations of p_b, p_s.

Proposition 2.11. *Assume (A1′), and let*

$$\Gamma(Q) := \inf_{\lambda>0} \frac{1}{\lambda}\left(V(\lambda Q|\mathcal{P}) - \inf_{\lambda'>0} \inf_{Q'\in\mathcal{M}_V} V(\lambda' Q'|\mathcal{P}) \right). \tag{2.26}$$

Then

$$p_b(B) = \inf_{Q\in\mathcal{M}_V} (E^Q[B] + \Gamma(Q)) \quad \forall B \in \mathcal{B}, \tag{2.27}$$

$$p_s(B) = \sup_{Q\in\mathcal{M}_V} (E^Q[B] - \Gamma(Q)) \quad \forall B \in -\mathcal{B}. \tag{2.28}$$

In particular, p_b (resp. p_s) is finite on $\mathcal{B}$ (resp. $-\mathcal{B}$), and if $B \in \mathcal{B} \cap (-\mathcal{B})$, then $p_s(B) = -p_b(-B)$.

Proof. By the argument above, we can apply Theorem 2.4 for all $x + B$. We prove only the case of p_b, but the same argument works also for p_s. By Theorem 2.4,

$$\sup_{\theta\in\Theta_{bb}} \inf_{P\in\mathcal{P}} E^P[U(\theta\cdot S_T - p + B)] = \inf_{\lambda>0} \inf_{Q\in\mathcal{M}_V} (V(\lambda Q|\mathcal{P}) - \lambda p + \lambda E^Q[B])$$

$$\sup_{\theta\in\Theta_{bb}} \inf_{P\in\mathcal{P}} E^P[U(\theta\cdot S_T)] = \inf_{\lambda>0} \inf_{Q\in\mathcal{M}_V} V(\lambda Q|\mathcal{P}) =: v_0.$$

Hence,

$$\begin{aligned}
&\sup_{\theta\in\Theta_{bb}} \inf_{P\in\mathcal{P}} E^P[U(\theta\cdot S_T - p + B)] \geq \sup_{\theta\in\Theta_{bb}} \inf_{P\in\mathcal{P}} E^P[U(\theta\cdot S_T)] \\
&\Leftrightarrow \inf_{\lambda>0} \inf_{Q\in\mathcal{M}_V} (V(\lambda Q|\mathcal{P}) - \lambda p + \lambda E^Q[B]) \geq v_0 \\
&\Leftrightarrow \inf_{Q\in\mathcal{M}_V} (V(\lambda Q|\mathcal{P}) + \lambda E^Q[B]) - \lambda p \geq v_0, \quad \forall \lambda > 0 \\
&\Leftrightarrow \inf_{Q\in\mathcal{M}_V} \left(\frac{1}{\lambda}(V(\lambda Q|\mathcal{P}) - v_0) + E^Q[B] \right) \geq p, \quad \forall \lambda > 0 \\
&\Leftrightarrow p \leq \inf_{Q\in\mathcal{M}_V} (E^Q[B] + \Gamma(Q)).
\end{aligned}$$

Therefore, we have (2.27). Also, $p_b(B) < \infty$ since $\Gamma(Q) \leq V(Q|\mathcal{P}) - v_0 < \infty$ and $B \in L^1(Q)$ for all $Q \in \mathcal{M}_V$. Finally, the next lemma shows that the set "$\{\cdots\}$" appearing in the definition of p_b is non-empty, hence $p_b(B) > -\infty$. □

Lemma 2.12. *For every $B \in \mathcal{B}$,*

$$\sup_{x\geq 0} \sup_{\theta\in\Theta_{bb}} \inf_{P\in\mathcal{P}} E^P[U(\theta\cdot S_T + x + B)] > \sup_{\theta\in\Theta_{bb}} \inf_{P\in\mathcal{P}} E^P[U(\theta\cdot S_T)] \tag{2.29}$$

Proof. By the concavity of U and the fact that Θ_{bb} is a cone, we have

$$\sup_{\theta\in\Theta_{bb}}\inf_{P\in\mathcal{P}} E^P[U(\theta\cdot S_T + x + B)] \geq \frac{\varepsilon}{1+\varepsilon}\sup_{\theta\in\Theta_{bb}}\inf_{P\in\mathcal{P}} E^P[U(\theta\cdot S_T)] + \frac{1}{1+\varepsilon}\inf_{P\in\mathcal{P}} E^P[U((1+\varepsilon)(x+B))],$$

where ε is taken so that (2.10) and (2.11) are satisfied. Thus, it suffices to see that

$$\sup_{x\geq 0}\inf_{P\in\mathcal{P}} E^P[U((1+\varepsilon)(x+B))] > \sup_{\theta\in\Theta_{bb}}\inf_{P\in\mathcal{P}} E^P[U(\theta\cdot S_T)]. \tag{2.30}$$

First, it is easy to deduce from (2.11) that $\{U((1+\varepsilon)(x+B))^- dP/d\mathbb{P}\}_{P\in\mathcal{P}}$ is uniformly integrable for every $x \geq 0$, hence $P \mapsto E^P[U((1+\varepsilon)(x+B))]$ is weakly lower semicontinuous on $\mathcal{P}$ (see the proof of Lemma 3.1). Also, noting that $B \in L^1(P)$ for all $P \in \mathcal{P}$ by (2.12) applied to $\nu = P$, $E^P[U((1+\varepsilon)(x+B))] \leq V(1) + (1+\varepsilon)E^P[x+B] < \infty$ for all $P \in \mathcal{P}$ and $x \geq 0$. Therefore, we can apply a minimax theorem as in the proof of Proposition 3.2 to obtain:

$$\begin{aligned}\sup_{x\geq 0}\inf_{P\in\mathcal{P}} E^P[U((1+\varepsilon)(x+B))] &= \inf_{P\in\mathcal{P}}\sup_{x\geq 0} E^P[U((1+\varepsilon)(x+B))]\\ &= \inf_{P\in\mathcal{P}}\lim_{x\nearrow\infty} E^P[U((1+\varepsilon)(x+B))]\\ &= \inf_{P\in\mathcal{P}} E^P\left[\lim_{x\nearrow\infty} U((1+\varepsilon)(x+B))\right]\\ &= U(\infty) = V(0).\end{aligned}$$

Here the third equality follows from the monotone convergence theorem. Finally, the RHS of (2.30) is written as $\inf_{\lambda>0}\inf_{Q\in\mathcal{M}_V} V(\lambda Q|\mathcal{P})$ by Theorem 2.4, which is strictly less than $V(0)$ (see the proof of Theorem 2.5 in Sect. 4.2). □

Let $(\hat{\lambda}_0\hat{Q}_0, \hat{P}_0)$ be a solution to the dual problem (2.15) with $B \equiv 0$, whose existence is guaranteed by Theorem 2.5.

Corollary 2.13. *Let $B, B' \in \mathcal{B}\cap(-\mathcal{B})$. Then*

(a) $p_b(0) = p_s(0) = 0.$
(b) $p_s(B+x) = p_s(B)+x,\ p_b(B+x) = p_b(B)+x,$ *for all* $x \in \mathbb{R}$.
(c) $B \leq B' \Rightarrow p_s(B) \leq p_s(B')$ *and* $p_b(B) \leq p_b(B')$.
(d) p_s *(resp.* p_b*) is a proper convex (resp. concave) function on* $-\mathcal{B}$ *(resp.* $\mathcal{B}$*);*
(e) $\inf_{Q\in\mathcal{M}_V^e} E^Q[B] \leq p_b(B) \leq E^{\hat{Q}_0}[B] \leq p_s(B) \leq \sup_{Q\in\mathcal{M}_V^e} E^Q[B].$

Proof. Noting that $\Gamma(Q) \geq 0$ for all $Q \in \mathcal{M}_V$, and $\min_{Q\in\mathcal{M}_V} \Gamma(Q) = \Gamma(\hat{Q}_0) = 0$, (a)–(d) are immediate from the risk measure expressions (2.27) and (2.28).

(e) Since $\Gamma(\hat{Q}_0) = 0$,

$$\begin{aligned}\inf_{Q\in\mathcal{M}_V} E^Q[B] &\leq \inf_{Q\in\mathcal{M}_V} (E^Q[B] + \Gamma(Q)) \\ &\leq E^{\hat{Q}_0}[B] + \Gamma(\hat{Q}_0) = E^{\hat{Q}_0}[B] \\ &= E^{\hat{Q}_0}[B] - \Gamma(\hat{Q}_0) \\ &\leq \sup_{Q\in\mathcal{M}_V} (E^Q[B] - \Gamma(Q)) \leq \sup_{Q\in\mathcal{M}_V} E^Q[B].\end{aligned}$$

It remains only that $\inf_{Q\in\mathcal{M}_V} E^Q[B] = \inf_{Q\in\mathcal{M}_V^e} E^Q[B]$, and $\sup_{Q\in\mathcal{M}_V} E^Q[B] = \sup_{Q\in\mathcal{M}_V^e} E^Q[B]$. Taking an element $\bar{Q}$ of $\mathcal{M}_V^e$, which is non-empty, we set $Q_\alpha := \alpha\bar{Q} + (1-\alpha)Q$ for each $Q \in \mathcal{M}_V$. Then $Q_\alpha \in \mathcal{M}_V^e$ for $\alpha \in (0, 1]$, and $\alpha \mapsto E^{Q_\alpha}[B]$ is continuous on $[0, 1]$, since $B \in L^1(Q)$ for all $Q \in \mathcal{M}_V$. This concludes the proof. □

Remark 2.14. The concept of utility indifference valuation is quite popular, which goes back to Hodges and Neuberger [18]. Especially, the case of exponential utility with risk measure representation has been extensively studied by many authors including Rouge and El Karoui [31], the "six-author paper" [10], Mania and Schweizer [24], among others. More recently, Owen and Žitković [29], Biagini et al. [6] consider the case of utility functions other than the exponential utility. Our Proposition 2.11 extends these existing results to the framework of *robust utility*. This extension is mathematically minor, and the only apparent difference emerges as the *robustification* of the penalty term. It proposes, however, a method of pricing contingent claims in the presence of model uncertainty.

In view of (a)–(d), p_b (resp. $B \mapsto p_s(-B)$) indeed defines a *concave monetary utility function* (resp. *convex risk measure*) on the convex set $\mathcal{B}$ in the sense of Biagini and Frittelli [5]. Although $\mathcal{B}$ is not a *linear space*, the concepts of convex risk measure and monetary utility function still make sense in view of the properties 1 and 2 of $\mathcal{B}$ listed just before Proposition 2.11.

Finally, $p_s(B)$ and $p_s(B)$ can be viewed as arbitrage-free prices in view of (e).

3. Duality

This section aims at proving the duality equality (Theorem 2.4). Our strategy is as follows. When $\mathcal{P}$ is a singleton, duality equalities are available in various settings. Thus, we first show that the order of "$\inf_{P\in\mathcal{P}}$" and "$\sup_{\theta\in\Theta_{bb}}$"

in the LHS of (2.17) are interchangeable. This reduces the *robust problem* to a family of *subjective problems*. Given this minimax equality, the duality will follow if we can apply the duality result of [27] to *all* $P \in \mathcal{P}$. Although this is not applicable to all $P \in \mathcal{P}$, we can take, in a rather trivial way, a "dense" subset of $\mathcal{P}$ on which the duality holds true. Then some approximation arguments conclude the proof. We begin with a simple lemma. Recall that we identify $\mathcal{P}$ with $\mathcal{D} := \{dP/d\mathbb{P} : P \in \mathcal{P}\} \subset L^1$ by the injection $P \mapsto dP/d\mathbb{P}$, and (A1)–(A5) are always assumed without particular mention.

Lemma 3.1. *The map $P \mapsto E^P[U(\theta \cdot S_T + B)]$ is weakly lower semicontinuous on $\mathcal{P}$ for each $\theta \in \Theta_{bb}$.*

Proof. Fix $\theta \in \Theta_{bb}$. By the definition of lower semicontinuity, it suffices to show that for any $\alpha \in \mathbb{R}$, the level set $A_\alpha := \{D \in \mathcal{D} : E[DU(\theta \cdot S_T + B)] \leq \alpha\}$ is weakly closed in L^1, and since A_α is clearly convex, we have only to show A_α is strongly closed. Let $(D_n)_n \subset A_\alpha$ be a convergent sequence, and $D = \lim_n D_n$ in L^1. Taking a subsequence, we may assume $D_n \to D$ a.s.

Since U is concave, and $\theta \cdot S_T$ is bounded from below by a (say),

$$D_n U(\theta \cdot S_T + B) \geq \frac{\varepsilon}{1+\varepsilon} D_n U\left(\frac{1+\varepsilon}{\varepsilon} a\right) + \frac{1}{1+\varepsilon} D_n U(-(1+\varepsilon)B^-)$$
$$=: G_n^\varepsilon.$$

The sequence $(G_n^\varepsilon)_n$ is a.s. convergent, and is uniformly integrable for some $\varepsilon > 0$ by (A5). Therefore, we can apply Fatou's lemma to conclude that

$$E[DU(\theta \cdot S_T + B)] \leq \liminf_{n\to\infty} E[D_n U(\theta \cdot S_T + B)] \leq \alpha,$$

hence $D \in A_\alpha$. □

Proposition 3.2. *We have*

$$\sup_{\theta\in\Theta_{bb}} \inf_{P\in\mathcal{P}} E^P[U(\theta \cdot S_T + B)] = \inf_{P\in\mathcal{P}} \sup_{\theta\in\Theta_{bb}} E^P[U(\theta \cdot S_T + B)]. \quad (3.1)$$

Proof. The map $P \mapsto E^P[U(\theta \cdot S_T + B)]$ is linear (hence convex), and weakly lower semicontinuous for every $\theta \in \Theta_{bb}$ by Lemma 3.1. Also, the set $\mathcal{P}$ is assumed to be weakly compact. On the other hand, $\theta \mapsto E^P[U(\theta \cdot S_T + B)]$ is concave since U is concave, and Θ_{bb} is convex. Therefore, we can apply Fan's minimax theorem (Simons [36], Theorem 3.2) to get (3.1). □

We proceed to the next step. By (A4) and Remark 2.2, we can take a pair $(Q_0, P_0) \in \mathcal{M}_V \times \mathcal{P}$ such that $Q_0 \sim P_0 \sim \mathbb{P}$ and $V(Q_0|P_0) < \infty$. Fixing such a pair, we set

$$P_\alpha := \alpha P + (1-\alpha)P_0, \quad \forall P \in \mathcal{P}.$$

Then define

$$\mathcal{P}_0 := \{P_\alpha : P \in \mathcal{P},\ \alpha \in [0,1)\}. \tag{3.2}$$

An important consequence of the assumption $U(\infty) < \infty$ in (A1′) is the next lemma, which is pointed out in Föllmer and Gundel [12], Remark 3.10 for a special choice of (Q_0, P_0), and the same proof applies to our case.

Lemma 3.3. *Assume (A1′). Then for every $P \in \mathcal{P}_0$, $V(Q_0|P) < \infty$.*

Note that (A5) implies in particular that

$$E^P[U(-\varepsilon B^+)] > -\infty \quad \text{and} \quad E^P[U(-(1+\varepsilon)B^-)] > -\infty, \qquad \forall P \in \mathcal{P}. \tag{3.3}$$

Also, by Lemma 3.3, we have

$$\mathcal{M}_V^e(P) \neq \emptyset, \quad \forall P \in \mathcal{P}_0. \tag{3.4}$$

Therefore, we can apply Theorem 2.1 of Owari [27] to get:

Lemma 3.4. *For every $P \in \mathcal{P}_0$, we have*

$$\sup_{\theta \in \Theta_{bb}} E^P[U(\theta \cdot S_T + B)] = \inf_{\lambda > 0} \inf_{Q \in \mathcal{M}_V(P)} (V(\lambda Q|P) + \lambda E^Q[B]). \tag{3.5}$$

In particular,

$$\inf_{P \in \mathcal{P}_0} \sup_{\theta \in \Theta_{bb}} E^P[U(\theta \cdot S_T + B)] = \inf_{\lambda > 0} \inf_{P \in \mathcal{P}_0} \inf_{Q \in \mathcal{M}_V(P)} (V(\lambda Q|P) + \lambda E^Q[B]). \tag{3.6}$$

Proof of Theorem 2.4. We first show:

$$\inf_{P \in \mathcal{P}} \sup_{\theta \in \Theta_{bb}} E^P[U(\theta \cdot S_T + B)] = \inf_{P \in \mathcal{P}_0} \sup_{\theta \in \Theta_{bb}} E^P[U(\theta \cdot S_T + B)]. \tag{3.7}$$

The inequality "$\leq$" is trivial since $\mathcal{P}_0 \subset \mathcal{P}$. To see the converse inequality, it suffices to check that the map $\alpha \mapsto \sup_{\theta \in \Theta_{bb}} E^{P_\alpha}[U(\theta \cdot S_T + B)]$ is upper semicontinuous on $[0,1]$. This map is convex as a pointwise supremum of affine functions, and finite valued since U is bounded from above by (A1′), hence upper semicontinuous.

Next we show that for all $\lambda > 0$,

$$\inf_{P \in \mathcal{P}_0} \inf_{Q \in \mathcal{M}_V(P)} (V(\lambda Q|P) + \lambda E^Q[B]) = \inf_{(Q,P) \in \mathcal{M}_V \times \mathcal{P}} (V(\lambda Q|P) + \lambda E^Q[B]). \tag{3.8}$$

Since $B \in L^1(Q)$ for all $Q \in \mathcal{M}_V$, $Q \in \mathcal{M}_V \setminus \mathcal{M}_V(P)$ implies $V(\lambda Q|P) + \lambda E^Q[B]) = +\infty$. Thus,

$$\inf_{P\in\mathcal{P}_0}\inf_{Q\in\mathcal{M}_V(P)}(V(\lambda Q|P)+\lambda E^Q[B])=\inf_{P\in\mathcal{P}_0}\inf_{Q\in\mathcal{M}_V}(V(\lambda Q|P)+\lambda E^Q[B]).$$

Now (3.8) follows by showing that for each $P\in\mathcal{P}$ and $\lambda>0$,

$$\inf_{Q\in\mathcal{M}_V}(V(\lambda Q|P)+\lambda E^Q[B])\geq\inf_{\alpha\in[0,1)}\inf_{Q\in\mathcal{M}_V}(V(\lambda Q|P_\alpha)+\lambda E^Q[B]). \tag{3.9}$$

If the LHS is $+\infty$, there is nothing to prove, hence we assume that the LHS is finite. Note that $P\mapsto\inf_{Q\in\mathcal{M}_V}(V(\lambda Q|P)+\lambda E^Q[B])$ is convex, since the $(Q,P)\mapsto V(\lambda Q|P)$ and the set $\mathcal{M}_V$ are convex. Therefore, $\alpha\mapsto\inf_{Q\in\mathcal{M}_V}(V(\lambda Q|P_\alpha)+\lambda E^Q[B])$ is convex, and finite on $[0,1]$ by assumption, hence upper semicontinuous. Thus (3.9) follows, and we get (3.8).

Finally, combining Proposition 3.2, Lemma 3.4, (3.7) and (3.8), we have

$$\begin{aligned}\sup_{\theta\in\Theta_{bb}}\inf_{P\in\mathcal{P}}E^P[U(\theta\cdot S_T+B)]&=\inf_{P\in\mathcal{P}_0}\sup_{\theta\in\Theta_{bb}}E^P[U(\theta\cdot S_T+B)]\\&=\inf_{\lambda>0}\inf_{P\in\mathcal{P}_0}\inf_{Q\in\mathcal{M}_V(P)}(V(\lambda Q|P)+\lambda E^Q[B])\\&=\inf_{\lambda>0}\inf_{(Q,P)\in\mathcal{M}_V\times\mathcal{P}}(V(\lambda Q|P)+\lambda E^Q[B]),\end{aligned}$$

and the proof is complete. □

Remark 3.5. Some related results are found. When the utility function is defined on $\mathbb{R}_+$ and $B\equiv0$, Schied and Wu [35] and Schied [34] proved similar duality results, on which our idea of proof is based. Wittmüss [37] dealt with the case of *bounded* endowment with utility function on $\mathbb{R}_+$. Our Theorem 2.4 is an extension of Theorem 2.4 of Owari [28] which investigated the case of exponential utility with unbounded endowment.

4. Solution to the dual problem

This section studies the dual problem (2.16). We begin with a brief review of a related result due to Föllmer and Gundel [12], explaining our idea and what our contribution is.

Suppose for a moment that $B\equiv0$, and consider the problem:

$$\text{minimize}\quad V(Q|\mathcal{P}),\quad\text{over } Q\in\mathcal{M}_V. \tag{4.1}$$

This is nothing other than the *robust f-projection* problem of [12] which shows the existence of a solution. The heart of the proof of [12] consists of getting the weak lower semicontinuity of the divergence $(Q,P)\mapsto V(Q|P)$, and a uniform integrability criterion in the spirit of the de la Vallée–Poussin criterion. These arguments can be modified, in a rather obvious way, to deal with the case where the V-divergence is penalized by $E^Q[B]$ with a *bounded* B.

When we pass from (4.1) to (2.16), two apparent differences arise. First, the domain is no longer a set of *probability* measures, but of *positive finite* measures, and the term $\nu(B)$ with *unbounded* B appears as a penalty function. The first point is not very problematic, and the crucial is the second. Actually, the penalty term $\nu \mapsto \nu(B)$ is no longer lower semicontinuous unless B is bounded from below. Thus, it is not *a priori* trivial whether $(\nu, P) \mapsto V(\nu|P) + \nu(B)$ has an appropriate lower semicontinuity. Also, we need an estimate between $V(\nu|\mathcal{P})$ and $V(\nu|\mathcal{P}) + \nu(B)$ to apply the uniform integrability criterion of [12].

In the case where U is the exponential utility, hence $V(\cdot|\cdot)$ is the relative entropy, Owari [28] considered the same problem, where (uniform) integrability conditions on B, corresponding to (A5) of our setup, clear everything. The proofs employed there rely on some explicit computation appealing to the nature of exponential functions, which is no longer possible in our general setup. However, we can fill the gap by some knowledge from elementary convex analysis.

We proceed as follows. After establishing some elementary estimates concerning the penalty term, we show that the penalized robust V-divergence functional $\nu \mapsto V(\nu|\mathcal{P})+\nu(B)$ is in fact weakly lower semicontinuous. Then the existence of a solution will follow from a similar argument with [12]. Again, recall that the assumptions (A1)–(A5) are in force throughout.

4.1. Preliminary estimates

Lemma 4.1. *There are real numbers c, C, as well as positive numbers c', C' depending only on $\mathcal{P}$, B and V such that*

$$c + c'V(\nu|\mathcal{P}) \le V(\nu|\mathcal{P}) + \nu(B) \le C + C'V(\nu|\mathcal{P}), \quad \forall \nu \in \mathrm{cone}(\mathcal{M}_V). \tag{4.2}$$

In particular,

$$\inf_{\nu\in\mathrm{cone}(\mathcal{M}_V)} (V(\nu|\mathcal{P}) + \nu(B)) \ge c + c'V(1) > -\infty. \tag{4.3}$$

Proof. Let $\nu \in \mathrm{cone}(\mathcal{M}_V)$, and take any $P \in \mathcal{P}$ such that $V(\nu|P) < \infty$. Then $\nu \ll P$, and $\nu(B)$ is finite. Fixing $\varepsilon > 0$ as in (A5), we have

$$\begin{aligned} -B\frac{d\nu}{dP} \le B^-\frac{d\nu}{dP} &= (1+\varepsilon)B^-\frac{1}{1+\varepsilon}\frac{d\nu}{dP} \\ &\le (1+\varepsilon)B^-\left\{\frac{1}{1+\varepsilon}\frac{d\nu}{dP} + \frac{\varepsilon}{1+\varepsilon}\right\} \end{aligned}$$

$$\leq V\left(\frac{1}{1+\varepsilon}\frac{d\nu}{dP}+\frac{\varepsilon}{1+\varepsilon}\right)-U(-(1+\varepsilon)B^-)$$
$$\leq \frac{1}{1+\varepsilon}V\left(\frac{d\nu}{dP}\right)+\frac{\varepsilon}{1+\varepsilon}V(1)-U(-(1+\varepsilon)B^-).$$

Taking P-expectation,

$$\begin{aligned}-\nu(B) &\leq \frac{1}{1+\varepsilon}V(\nu|P)+\frac{\varepsilon}{1+\varepsilon}V(1)-E^P[U(-(1+\varepsilon)B^-)]\\ &\leq \frac{1}{1+\varepsilon}V(\nu|P)+\frac{\varepsilon}{1+\varepsilon}V(1)-\inf_{P'\in\mathcal{P}}E^{P'}[U(-(1+\varepsilon)B^-)].\end{aligned}$$

The last term in the second line is finite by (A5), and this holds for every $P\in\mathcal{P}$, since the last side is $+\infty$ if $V(\nu|P)=\infty$. Therefore, taking the infimum over $\mathcal{P}$, we get

$$V(\nu|\mathcal{P})+\nu(B)\geq \frac{\varepsilon}{1+\varepsilon}V(\nu|\mathcal{P})-\frac{\varepsilon}{1+\varepsilon}V(1)+\inf_{P\in\mathcal{P}}E^P[U(-(1+\varepsilon)B^-)].$$

This shows the first inequality in (4.2) with

$$c=-\frac{\varepsilon}{1+\varepsilon}V(1)+\inf_{P\in\mathcal{P}}E^P[U(-(1+\varepsilon)B^-)],\quad c'=\frac{\varepsilon}{1+\varepsilon}.$$

Similarly, for any $P\in\mathcal{P}$,

$$\begin{aligned}\varepsilon\nu(B)\leq\varepsilon\nu(B^+) &\leq V(\nu|P)-E^P[U(-\varepsilon B^+)]\\ &\leq V(\nu|P)-\inf_{P'\in\mathcal{P}}E^{P'}[U(-\varepsilon B^+)].\end{aligned}$$

This implies

$$V(\nu|\mathcal{P})+\nu(B)\leq\frac{1+\varepsilon}{\varepsilon}V(\nu|\mathcal{P})-\frac{1}{\varepsilon}\inf_{P\in\mathcal{P}}E^P[U(-\varepsilon B^+)].$$

We thus obtain the second inequality in (4.2) with $C=-\inf_{P\in\mathcal{P}}E^P[U(-\varepsilon B^+)]/\varepsilon$ and $C'=(1+\varepsilon)/\varepsilon$. □

Let $\mathbb{M}$ denote the space of positive finite measures on $(\Omega,\mathcal{F})$ absolutely continuous w.r.t. $\mathbb{P}$, which we identify with L^1_+ by the injection $\nu\mapsto d\nu/d\mathbb{P}$. It is known that the map $(\nu,P)\mapsto V(\nu|P)$ is weakly lower semicontinuous on $\mathbb{M}^2$ (see [12, Lemma 2.7]). Further, since $\mathcal{P}$ is weakly compact, we have even that $\nu\mapsto V(\nu|\mathcal{P})$ is weakly lower semicontinuous on $\mathbb{M}$. The next lemma extends this to the penalized robust V-divergence functional $\nu\mapsto V(\nu|\mathcal{P})+\nu(B)$. Note that $\text{cone}(\mathcal{M}_{loc})=\{\lambda Q:\lambda\geq 0,\ Q\in\mathcal{M}_{loc}\}$ is closed, hence the set $\{\nu\in\text{cone}(\mathcal{M}_{loc}):V(\nu|\mathcal{P})\leq\alpha\}$ is a closed convex set for each $\alpha\in\mathbb{R}$.

Lemma 4.2. *The functional* $\nu \mapsto V(\nu|\mathcal{P}) + \nu(B)$ *is weakly lower semicontinuous on* $\mathrm{cone}(\mathcal{M}_V)$*, that is, for every* $\alpha \in \mathbb{R}$*, the level set*

$$A_\alpha := \{\nu \in \mathrm{cone}(\mathcal{M}_V) : V(\nu|\mathcal{P}) + \nu(B) \leq \alpha\}$$

is weakly closed in L^1.

Note that the convex set $\mathrm{cone}(\mathcal{M}_V)$ is not closed in general. But this will not cause any problem. The lemma states that if we extend the functional $V(\nu|\mathcal{P}) + \nu(B)$ to the whole $\mathbb{M}$ by setting $+\infty$ outside $\mathrm{cone}(\mathcal{M}_V)$, then it is lower semicontinuous in the usual sense.

Proof. Since $\nu \mapsto V(\nu|\mathcal{P}) + \nu(B)$ is convex, it suffices to show that the level set A_α is strongly closed for each α. Let $(\nu_n)_n$ be a Cauchy sequence in A_α with the limit $\nu \in \mathbb{M}$. We first note that $\nu \in \mathrm{cone}(\mathcal{M}_V)$. Indeed, by Lemma 4.1, $V(\nu_n|\mathcal{P}) + \nu_n(B) \leq \alpha$ implies $\nu_n \in \{\nu' \in \mathrm{cone}(\mathcal{M}_{loc}) : V(\nu'|\mathcal{P}) \leq (\alpha - c)/c'\}$ which is a closed subset of $\mathrm{cone}(\mathcal{M}_V)$. Thus $\nu(B)$ is finite and $V(\nu|\mathcal{P}) + \nu(B)$ is well-defined. Also, by taking subsequence, we may assume that $d\nu_n/d\mathbb{P} \to d\nu/d\mathbb{P}$, a.s.

Since $\mathcal{P}$ is assumed to be weakly compact, there exists for each n a $P_n \in \mathcal{P}$ such that $V(\nu_n|\mathcal{P}) = V(\nu_n|P_n)$. Then the Komlós theorem (Delbaen and Schachermayer [9], Theorem 1.3) shows the existence of another sequence $\{(\tilde{\nu}_n, \tilde{P}_n)\}_n$ such that

$$(\tilde{\nu}_n, \tilde{P}_n) \in \mathrm{conv}\{(\nu_n, P_n), (\nu_{n+1}, P_{n+1}), ...\},$$

and $d\tilde{P}_n/d\mathbb{P} \to D$, a.s. for some positive random variable D, and again since $\mathcal{P}$ is weakly compact ($\Leftrightarrow$ closed and uniformly integrable by Lemma 2.1), the convergence takes place in L^1, hence $D = dP/d\mathbb{P}$ with $P \in \mathcal{P}$.

Noting that $\tilde{\nu}_n \ll \tilde{P}_n$ since $V(\tilde{\nu}_n|\tilde{P}_n) \leq (\alpha - c)/c' < \infty$, we have for each n,

$$-\frac{d\tilde{\nu}_n}{d\tilde{P}_n}B \leq \frac{d\tilde{\nu}_n}{d\tilde{P}_n}B^- \leq V\left(\frac{d\tilde{\nu}_n}{d\tilde{P}_n}\right) - U(-B^-),$$

hence multiplying both sides by $d\tilde{P}_n/d\mathbb{P}$ and rearranging the terms,

$$V\left(\frac{d\tilde{\nu}_n}{d\mathbb{P}}, \frac{d\tilde{P}_n}{d\mathbb{P}}\right) + \frac{d\tilde{\nu}_n}{d\mathbb{P}}B \geq \frac{d\tilde{P}_n}{d\mathbb{P}}U(-B^-),$$

for each n. Since $V(\cdot, \cdot)$ is lower semicontinuous,

$$V\left(\frac{d\nu}{d\mathbb{P}}, \frac{dP}{d\mathbb{P}}\right) + \frac{d\nu}{d\mathbb{P}}B \leq \liminf_{n\to\infty}\left(V\left(\frac{d\tilde{\nu}_n}{d\mathbb{P}}, \frac{d\tilde{P}_n}{d\mathbb{P}}\right) + \frac{d\tilde{\nu}_n}{d\mathbb{P}}B\right), \quad \text{a.s.}$$

Also, $\{(d\tilde{P}_n/d\mathbb{P})U(-B^-)\}_n$ is uniformly integrable by (A5) and the monotonicity of U, and converges a.s. to $(dP/d\mathbb{P})U(-B^-)$ by construction. Therefore, we can apply Fatou's lemma to conclude:

$$V(\nu|\mathcal{P}) + \nu(B) \leq V(\nu|P) + \nu(B) \leq \liminf_{n\to\infty}(V(\nu_n|P_n) + \nu_n(B)) \leq \alpha.$$

Thus, we have $\nu \in A_\alpha$, and the assertion is proved. □

Recall that $\lim_{x\to\infty} V(x)/x = +\infty$ by (A1). Then the next one is a restatement of Föllmer and Gundel [12], Lemma 2.12. Although the original version in [12] is stated with $\mathcal{Q}$ consisting of *probability* measures since their primal interest is in the problem (4.1), the *exactly* same proof applies to the case where $\mathcal{Q}$ is a set of *positive finite* measures.

Lemma 4.3. *Let $\mathcal{Q}$ be a set of positive finite measures. If*

$$\sup_{\nu\in\mathcal{Q}} V(\nu|\mathcal{P}) < \infty, \tag{4.4}$$

then $\mathcal{Q}$ is uniformly integrable as a subset of L^1, hence weakly relatively compact.

Proof. We build a bridge between [12, Lemma 2.12] and the current assertion. Since $\lim_{x\to\infty} V(x)/x = +\infty$ and $\mathcal{P}$ is weakly compact by (A3), [12, Lemma 2.12] together with the comment above shows the existence of a function $l : [0,\infty) \to [0,\infty)$ such that $\lim_{x\to\infty} l(x)/x = +\infty$ and

$$\forall c > 0,\ \exists c_0 > 0 \text{ s.t. } V(\nu|\mathcal{P}) \leq c \Rightarrow E[l(d\nu/d\mathbb{P})] \leq c_0.$$

Taking $c = \sup_{\nu\in\mathcal{Q}} V(\nu|\mathcal{P})$, we have $\sup_{\nu\in\mathcal{Q}} E[l(d\nu/d\mathbb{P})] < \infty$. The assertion then follows from the de la Vallée–Poussin criterion [11, Theorem II.22]. □

4.2. Existence of a solution

Proof of Theorem 2.5 (existence). We first prove that the problem

$$\text{minimize} \quad V(\nu|\mathcal{P}) + \nu(B), \quad \text{over } \nu \in \text{cone}(\mathcal{M}_V) \tag{4.5}$$

admits a minimizer, and then we verify that the minimizer must not be zero.

Let $\alpha \in \mathbb{R}$ be such that the level set $A_\alpha := \{\nu \in \text{cone}(\mathcal{M}_V) : V(\nu|\mathcal{P}) + \nu(B) \leq \alpha\}$ is non-empty (such α exists by (A4)). Then Lemma 4.1 shows that A_α is contained in the set:

$$\{\nu \in \text{cone}(\mathcal{M}_V) : V(\nu|\mathcal{P}) \leq \alpha'\},$$

for some α', which is uniformly integrable by Lemma 4.3. Therefore, A_α is weakly compact, and $\nu \mapsto V(\nu|\mathcal{P}) + \nu(B)$ is weakly lower semicontinuous on A_α, by Lemma 4.2. We thus obtain the existence of a minimizer $\hat{\nu} \in \text{cone}(\mathcal{M}_V)$.

It remains to verify that $\hat{\nu} \neq 0$. We may assume $V(0) < \infty$ since otherwise the assertion is trivial. To see this, it suffices to find an element $(\bar{\nu}, \bar{P})$ of $\text{cone}(\mathcal{M}_V) \times \mathcal{P}$ with $\bar{\nu} \sim \bar{P} \sim \mathbb{P}$ and $V(\bar{\nu}|\bar{P}) + \bar{\nu}(B) < V(0)$. Let $(\bar{Q}, \bar{P}) \in \mathcal{M}_V \times \mathcal{P}$ be such that $\bar{Q} \sim \bar{P}$ and $V(\bar{Q}|\bar{P}) < \infty$, whose existence is guaranteed by (A4) and Remark 2.2, and set $\nu_\lambda = \lambda\bar{Q}$. Consider the random map $\lambda \mapsto G(\lambda) := V(\lambda d\bar{Q}/d\bar{P}) + \lambda(d\bar{Q}/d\bar{P})B$, which is a.s. convex, hence $(G(\lambda) - G(0))/\lambda$ is increasing in λ, and $E^{\bar{P}}[G(\lambda)] = V(\lambda\bar{Q}|\bar{P}) + \lambda E^{\bar{Q}}[B]$. Then the monotone convergence theorem shows:

$$\lim_{\lambda \searrow 0} E^{\bar{P}}\left[\frac{G(\lambda) - G(0)}{\lambda}\right] = V'(0) + E^{\bar{Q}}[B] = -\infty.$$

This implies that $V(\nu_\lambda|\bar{P}) + \nu_\lambda(B) < V(0)$ for small enough $\lambda > 0$. □

Proof of Theorem 2.5 (construction of maximal solution). Let $\mathcal{S}$ be the set of all solutions to (2.16). Letting $\alpha = \inf_{\nu \in \text{cone}(\mathcal{M}_V)}(V(\nu|\mathcal{P}) + \nu(B))$, this set is written as:

$$\mathcal{S} := \{\nu \in \text{cone}(\mathcal{M}_V) : V(\nu|\mathcal{P}) + \nu(B) \leq \alpha\}. \tag{4.6}$$

In particular, $\mathcal{S}$ is a weakly compact convex set. Therefore, we see that $\mathcal{S}$ is even *countably convex*, i.e., if $(\nu^n)_n \subset \mathcal{S}$, and $\{\alpha_n\} \subset \mathbb{R}_+$ with $\sum_n \alpha_n = 1$, then $\sum_n \alpha_n \nu^n \in \mathcal{S}$.

Next, define

$$\mathcal{U} := \{\{d\nu/d\mathbb{P} > 0\} : \nu \in \mathcal{S}\}.$$

By the countable convexity of $\mathcal{S}$, the set $\mathcal{U}$ is σ-additive. Indeed, if $A_n \in \mathcal{U}$ and ν^n is a corresponding element of $\mathcal{S}$ for each n, the countable convexity of $\mathcal{S}$ implies that the measure $\nu = \sum_n 2^{-n}\nu^n$ is in $\mathcal{S}$, thus

$$\bigcup_n A_n = \bigcup_n \{d\nu^n/d\mathbb{P} > 0\} = \{d\nu/d\mathbb{P} > 0\} \in \mathcal{U}.$$

Now let $\{A_n\}$ be a sequence in $\mathcal{U}$ such that

$$\mathbb{P}(A_n) \nearrow \sup\{\mathbb{P}(A) : A \in \mathcal{U}\}.$$

Replacing A_n by $A_1 \cup \cdots \cup A_n$, we may assume that $A_n \subset A_{n+1}$ for each n. Define $\hat{A} := \cup_n A_n$, which belongs to $\mathcal{U}$ by countable additivity. Then

$$\mathbb{P}(\hat{A}) = \lim_{n\to\infty} \mathbb{P}(A_n) = \sup\{\mathbb{P}(A) : A \in \mathcal{U}\}.$$

Finally, an element $\hat{\nu} \in \mathcal{S}$ corresponding to $\hat{A}$ is a desired maximal solution. Indeed, if this is not maximal, then there exists some $\nu \in \mathcal{S}$ with $\nu \not\ll \hat{\nu}$, or equivalently $\mathbb{P}(A \setminus \hat{A}) > 0$ where $A = \{d\nu/d\mathbb{P} > 0\}$. Then

$$\mathbb{P}(A \cup \hat{A}) = \mathbb{P}(\hat{A}) + \mathbb{P}(A \setminus \hat{A}) > \mathbb{P}(\hat{A}).$$

This contradicts to the definition of $\hat{A}$, hence $\hat{\nu}$ must be maximal. □

4.3. Variational characterizations

Let us fix a solution $(\hat{\nu}, \hat{P}) = (\hat{\lambda}\hat{Q}, \hat{P}) \in (\mathrm{cone}(\mathcal{M}_V) \setminus \{0\}) \times \mathcal{P}$ to the dual problem. We characterize this solution via variational inequalities.

For any $Q \in \mathcal{M}_V$ and $\alpha \in [0, 1]$, we set

$$Q_\alpha := \alpha Q + (1 - \alpha)\hat{Q}.$$

Similar notations apply to $\nu \in \mathrm{cone}(\mathcal{M}_V)$ and $P \in \mathcal{P}$ in an obvious way.

Theorem 4.4. *Let $(\hat{\nu}, \hat{P}) = (\hat{\lambda}\hat{Q}, \hat{P})$ be as above.*

(a) If $Q \in \mathcal{M}_V$ satisfies $V(Q_{\alpha_0}|\hat{P}) < \infty$ for some $\alpha_0 \in (0, 1]$, $V'(\hat{\lambda}d\hat{Q}/d\hat{P}) + B \in L^1(Q)$, and

$$0 = E^{\hat{Q}}\left[V'\left(\hat{\lambda}\frac{d\hat{Q}}{d\hat{P}}\right) + B\right] \le E^{Q}\left[V'\left(\hat{\lambda}\frac{d\hat{Q}}{d\hat{P}}\right) + B\right]. \tag{4.7}$$

(b) If $P \in \mathcal{P}$ satisfies $V(\hat{Q}|P_{\alpha_0}) < \infty$ for some $\alpha_0 \in (0, 1]$, we have $V(\hat{\lambda}d\hat{Q}/d\hat{P}) - \hat{\lambda}(d\hat{Q}/d\hat{P})V'(\hat{\lambda}d\hat{Q}/d\hat{P}) \in L^1(P)$, and (with the convention $0 \cdot \infty = 0$),

$$\begin{aligned} E^{\hat{P}}&\left[V\left(\hat{\lambda}\frac{d\hat{Q}}{d\hat{P}}\right) - \hat{\lambda}\frac{d\hat{Q}}{d\hat{P}}V'\left(\hat{\lambda}\frac{d\hat{Q}}{d\hat{P}}\right)\right] \\ &\le E^{P}\left[V\left(\hat{\lambda}\frac{d\hat{Q}}{d\hat{P}}\right) - \hat{\lambda}\frac{d\hat{Q}}{d\hat{P}}V'\left(\hat{\lambda}\frac{d\hat{Q}}{d\hat{P}}\right)\right] \\ &\quad + V(0)P\left(\frac{d\hat{P}}{d\mathbb{P}} = 0\right). \end{aligned} \tag{4.8}$$

Here the second term of the RHS does not appear if either $V(0) = +\infty$ or $P \ll \hat{P}$.

(c) $\hat{Q} \sim \hat{P}$.

We first derive an auxiliary variational inequality from which Theorem 4.4 follows easily.

Proposition 4.5. *Let* $(\nu, P) \in (\mathrm{cone}(\mathcal{M}_V) \setminus \{0\}) \times \mathcal{P}$ *with* $V(\nu_{\alpha_0}|P_{\alpha_0}) < \infty$ *for some* $\alpha_0 \in (0, 1)$.

(a) Let

$$\begin{aligned}\Xi(\nu, P) := {} & 1_{\{\frac{d\hat{P}}{d\mathbb{P}}>0\}} \left\{ V'\left(\frac{d\hat{\nu}}{d\hat{P}}\right) + B \right\} \left(\frac{d\nu}{d\mathbb{P}} - \frac{d\hat{\nu}}{d\mathbb{P}} \right) \\ & + 1_{\{\frac{d\hat{P}}{d\mathbb{P}}>0\}} \left\{ V\left(\frac{d\hat{\nu}}{d\hat{P}}\right) - \frac{d\hat{\nu}}{d\hat{P}} V'\left(\frac{d\hat{\nu}}{d\hat{P}}\right) \right\} \left(\frac{dP}{d\mathbb{P}} - \frac{d\hat{P}}{d\mathbb{P}} \right) \\ & + 1_{\{\frac{d\hat{P}}{d\mathbb{P}}=0,\, \frac{dP}{d\mathbb{P}}>0\}} \left\{ V\left(\frac{d\nu}{d\mathbb{P}}, \frac{dP}{d\mathbb{P}}\right) + \frac{d\nu}{d\mathbb{P}} B \right\}. \end{aligned} \tag{4.9}$$

Then $\Xi(\nu, P) \in L^1(\mathbb{P})$ *and*

$$E[\Xi(\nu, P)] \geq 0. \tag{4.10}$$

(b) We have

$$\mathbb{P}\left(\frac{d\hat{P}}{d\mathbb{P}} > 0, \frac{d\hat{\nu}}{d\mathbb{P}} = 0, \frac{d\nu}{d\mathbb{P}} > 0 \right) = 0. \tag{4.11}$$

Proof. Fixing (ν, P) as above, define

$$G(\alpha, \nu, P) := V(d\nu_\alpha/d\mathbb{P}, dP_\alpha/d\mathbb{P}) + (d\nu_\alpha/d\mathbb{P})B, \quad \alpha \in [0, 1).$$

Since $(\nu, P) \mapsto V(d\nu/d\mathbb{P}, dP/d\mathbb{P})$ is convex, the map $\alpha \mapsto G(\alpha, \nu, P)$ is convex, a.s., hence $\alpha \mapsto \Xi(\alpha; \nu, P) := (G(\alpha, \nu, P) - G(0, \nu, P))/\alpha$ decreases a.s. to a limit $\Xi(\nu, P)$ as α tends to 0. Noting that $\Xi(\nu, P) \leq \Xi(\alpha_0, \nu, P) \in L^1$ by assumption, thus $\Xi(\nu, P)^+ \in L^1$, the dominated convergence theorem applied to the non-negative increasing sequence $\{\Xi(\alpha_0, \nu, P) - \Xi(\alpha, \nu, P)\}_\alpha$ shows:

$$E[\Xi(\nu, P)] = \lim_{\alpha \searrow 0} E[\Xi(\alpha; \nu, P)] \geq 0.$$

Here the last inequality follows from the fact:

$$E[\Xi(\alpha, \nu, P)] = \frac{V(\nu_\alpha|P_\alpha) + \nu_\alpha(B) - \left(V(\hat{\nu}|\hat{P}) + \hat{\nu}(B)\right)}{\alpha} \geq 0.$$

Thus we have $\Xi(\nu, P) \in L^1$, and the latter assertion in part (a) follows.

It remains to compute $\Xi(\nu, P)$ explicitly. To facilitate the notations, we denote $Z := d\nu/d\mathbb{P}$, $\hat{Z} := d\hat{\nu}/d\mathbb{P}$, $Z_\alpha := d\nu_\alpha/d\mathbb{P}$, $D := dP/d\mathbb{P}$, $\hat{D} := d\hat{P}/d\mathbb{P}$, and $D_\alpha := dP_\alpha/d\mathbb{P}$. Note first that $\Xi(\nu, P) = 0$ on the set $\{\hat{D} = D = 0\}$, since then $Z = \hat{Z} = 0$, hence $G(\alpha) \equiv 0$.

On $\{\hat{D} = 0, D > 0\}$, we have $\hat{Z} = 0$, thus $D\alpha = \alpha D$ and $Z_\alpha = \alpha Z$. Therefore, $G(\alpha) = V(\alpha Z, \alpha D) + \alpha ZB = \alpha DV(Z/D) + \alpha ZB$, hence

$$\Xi(\nu, P) = V(Z, D) + ZB = DV(Z/D) + ZB,$$

while on $\{\hat{D} > 0, \hat{Z} > 0\}$,

$$\begin{aligned}\Xi(\nu, P) = &\left\{V'(\hat{Z}/\hat{D})) + B\right\}\left(Z - \hat{Z}\right) \\ &+ \left\{V(\hat{Z}/\hat{D}) - (\hat{Z}/\hat{D})V'(\hat{Z}/\hat{D})\right\}\left(D - \hat{D}\right).\end{aligned}$$

The case of $\{\hat{D} > 0, \hat{Z} = 0\}$ needs more care. Note that this set has $\mathbb{P}$-probability zero if $V(0) = +\infty$, hence we assume $V(0) < \infty$. Then

$$\begin{aligned}\Xi(\nu, P) &= \lim_{\alpha \searrow 0} \frac{V(\alpha Z, D_\alpha) - V(0, \hat{D})}{\alpha} + ZB \\ &= \lim_{\alpha \searrow 0} D_\alpha \frac{V(\alpha Z/D_\alpha) - V(0)}{\alpha} + V(0) \lim_{\alpha \searrow 0} \frac{D_\alpha - \hat{D}}{\alpha} + ZB \\ &= \lim_{\alpha \searrow 0} Z \frac{V(\alpha Z/D_\alpha) - V(0)}{\alpha Z/D_\alpha} + V(0)(D - \hat{D}) + ZB \\ &= ZV'(0) + V(0)(D - \hat{D}) + ZB.\end{aligned}$$

Summing up the terms, we have (a). Also, since $V'(0) = -\infty$,

$$\Xi(\nu, P) = -\infty \quad \text{on} \quad \left\{\frac{d\hat{P}}{d\mathbb{P}} > 0, \ \frac{d\hat{\nu}}{d\mathbb{P}} = 0, \ \frac{d\nu}{d\mathbb{P}} > 0\right\}.$$

If this set has positive probability, then we have a contradiction to the fact that $\Xi(\nu, P) \in L^1(\mathbb{P})$, hence (b). □

Proof of Theorem 4.4. (c) By (A4) and Remark 2.2, there exists a pair $(\bar{Q}, \bar{P}) \in \mathcal{M}_V \times \mathcal{P}$ such that $\bar{Q} \sim \bar{P} \sim \mathbb{P}$ and $V(\bar{Q}|\bar{P}) < \infty$, which satisfies the assumption of Proposition 4.5. Since $\bar{Q} \sim \mathbb{P}$, Proposition 4.5 (b) yields

$$\mathbb{P}\left(\frac{d\hat{P}}{d\mathbb{P}} > 0, \ \frac{d\hat{Q}}{d\mathbb{P}} = 0\right) = 0.$$

This implies $\hat{P} \ll \hat{Q}$, hence $\hat{P} \sim \hat{Q}$.

(a) Let $P = \hat{P}$, then $P_{\alpha_0} = \hat{P}$. Note first that $(d\hat{\nu}/d\hat{P})|V'(d\hat{\nu}/d\hat{P})| \leq C_1' V(d\hat{\nu}/d\hat{P}) + C_2'(d\hat{\nu}/d\hat{P} + 1) \in L^1(\hat{P})$ by (2.6) ($\Leftrightarrow$ (A2)), hence $V'(\hat{\lambda} d\hat{Q}/d\hat{P}) + B = V'(d\hat{\nu}/d\hat{P}) + B \in L^1(\hat{Q})$. For any $\nu = \lambda Q \in \mathrm{cone}(\mathcal{M}_V)\setminus\{0\}$ with $V(\nu_\alpha|\hat{P}) < \infty$ for some $\alpha \in (0,1)$, Proposition 4.5 then shows that $V'(\hat{\lambda} d\hat{Q}/d\hat{P}) + B \in L^1(\nu) = L^1(Q)$ and

$$\hat{\nu}\left(V'\left(\hat{\lambda}\frac{d\hat{Q}}{d\hat{P}}\right) + B\right) \leq \nu\left(V'\left(\hat{\lambda}\frac{d\hat{Q}}{d\hat{P}}\right) + B\right). \tag{4.12}$$

Taking $\nu = \hat{\lambda} Q$ with $V(Q_\alpha|\hat{P}) < \infty$ for some $\alpha \in (0,1)$, we have the second inequality in (4.7). On the other hand, every $\nu = \lambda \hat{Q}$ with $\lambda > 0$ satisfies $V(\nu_\alpha|\hat{P}) < \infty$, hence

$$E^{\hat{Q}}\left[V'\left(\hat{\lambda}\frac{d\hat{Q}}{d\hat{P}}\right) + B\right] \leq \lambda E^{\hat{Q}}\left[V'\left(\hat{\lambda}\frac{d\hat{Q}}{d\hat{P}}\right) + B\right], \quad \forall \lambda > 0.$$

This is possible only if the LHS is zero, thus we obtain the first equality in (4.7).

(b) Similarly, we set $\nu = \hat{\nu} = \hat{\lambda}\hat{Q}$. Then noting that $V(0, y) = yV(0)$ by (2.7), and $\nu = \hat{\nu} \ll \hat{P}$, (4.9) is written (with the convention $0 \cdot \infty = 0$) as

$$\begin{aligned}\Xi(\hat{\nu}, P) = &\left\{V\left(\hat{\lambda}\frac{d\hat{Q}}{d\hat{P}}\right) - \hat{\lambda}\frac{d\hat{Q}}{d\hat{P}} V'\left(\hat{\lambda}\frac{d\hat{Q}}{d\hat{P}}\right)\right\}\left(\frac{dP}{d\mathbb{P}} - \frac{d\hat{P}}{d\mathbb{P}}\right) \\ &+ \frac{dP}{d\mathbb{P}} V(0) 1_{\{\frac{d\hat{P}}{d\mathbb{P}} = 0, \frac{dP}{d\mathbb{P}} > 0\}}.\end{aligned}$$

Here the set $\{\frac{d\hat{P}}{d\mathbb{P}} = 0, \frac{dP}{d\mathbb{P}} > 0\}$ is a null set if $P \ll \hat{P}$, and note that when $V(0) = +\infty$, the assumption $V(\hat{Q}|P_\alpha) < \infty$ implies $\hat{Q} \sim P_\alpha$, hence $P \ll P_\alpha \sim \hat{Q} \ll \hat{P}$. The assertion now follows by taking the expectation of $\Xi(\hat{\nu}, P)$. □

Proof of Theorem 2.7. The assertion (a) is nothing other than Theorem 4.4 (c).

(b) Given the variational inequality (4.7), the representation (2.19) follows from a standard argument using a version of Hahn–Banach theorem and Yor's closedness theorem as Goll and Rüschendorf [15], Theorems 3.2 and 7.1. However, we give a proof for the convenience of the reader.

Let $\mathcal{L}$ be the vector space of all variables of the form $f = \theta \cdot S_T$ where θ runs through all $(S, \hat{Q})$-integrable processes such that $\theta_0 = 0$ and each $\theta \cdot S$ is a $\hat{Q}$-martingale (not only local). It suffices to show that $\varphi := V'(\hat{\lambda} d\hat{Q}/d\hat{P}) + \hat{\lambda}(d\hat{Q}/d\hat{P})B \in \mathbb{R} \oplus \mathcal{L}$ where "$\oplus$" denotes the direct sum, since this already implies $\varphi \in \mathcal{L}$ in view of the first equality in (4.7).

Now Yor's theorem ([38, Corollaire 2.5.2], [9, Theorem 1.6]) shows that $\mathcal{L}$ is closed in $L^1(\hat{Q})$, hence so is $\mathbb{R} \oplus \mathcal{L}$. Thus, if $\varphi \notin \mathbb{R} \oplus \mathcal{L}$, the Hahn–Banach theorem shows the existence of an $h \in L^\infty$ with $\|h\|_\infty = 1/2$ such that $E^{\hat{Q}}[h\varphi] > 0$, $E^{\hat{Q}}[h] = 0$, and $E^{\hat{Q}}[hf] = 0$ for all $f \in \mathcal{L}$.

Define a new probability $\tilde{Q}$ by setting $d\tilde{Q}/d\hat{Q} := 1 - h$. Noting that $d\tilde{Q}/d\hat{P} = (1-h)d\hat{Q}/d\hat{P}$ and $(1-h) \in [1/2, 3/2]$ by construction, we have $V(\tilde{Q}|\hat{P}) < \infty$ by (2.5). Also, $E^{\tilde{Q}}[f] = E^{\hat{Q}}[f] - E^{\hat{Q}}[hf] = 0$ for all $f \in \mathcal{L}$, which implies that $\tilde{Q}$ is a local martingale measure, hence $\tilde{Q} \in \mathcal{M}_V$. Finally,

$$E^{\tilde{Q}}[\varphi] = E^{\hat{Q}}[\varphi] - E^{\hat{Q}}[h\varphi] < E^{\hat{Q}}[\varphi].$$

This contradicts to (4.7), hence we must have $\varphi \in \mathbb{R} \oplus \mathcal{L}$. □

Remark 4.6. When $B \equiv 0$, parts (a) and (b) of Theorem 4.4 are contained in [12, Lemma 3.12], although an additional assumption is required for part (c). Since $\hat{Q}$ is a minimizer of $Q \mapsto V(\hat{\lambda} Q|\hat{P}) + \hat{\lambda} E^Q[B]$, the variational inequality (4.7) follows essentially from[15, Proposition 7.2]. A similar remark applies also to (4.8). However, our auxiliary characterization (Proposition 4.5) allows us to prove all at once without any additional assumption for the equivalence $\hat{Q} \sim \hat{P}$.

We conclude this section by proving the fact noted in the comment after Theorem 2.5.

Proposition 4.7. *Let* (ν_0, P_0) *and* (ν_1, P_1) *be two solutions to the dual problem, which have the maximal support among all solutions. Then we have*

$$\frac{d\nu_0}{dP_0} = \frac{d\nu_1}{dP_1}, \quad a.s. \tag{4.13}$$

Proof. Since (ν_0, P_0) and (ν_1, P_1) are solutions, $(\nu_\alpha, P_\alpha) := (\alpha\nu_1 + (1-\alpha)\nu_0, \alpha P_1 + (1-\alpha)P_0)$ is also a solution for every $\alpha \in (0, 1)$. Therefore,

$$\alpha(V(\nu_1|P_1) + \nu_1(B)) + (1-\alpha)(V(\nu_0|P_0) + \nu_0(B)) = V(\nu_\alpha|P_\alpha) + \nu_\alpha(B). \tag{4.14}$$

Since $\nu_0 \sim \nu_1 \sim P_0 \sim P_1 \sim P_\alpha$ by Theorem 4.4 and the maximality of solutions, the LHS is written as

$$E^{P_\alpha}\left[\alpha \frac{dP_1}{dP_\alpha} V\left(\frac{d\nu_1}{dP_1}\right) + (1-\alpha)\frac{dP_0}{dP_\alpha} V\left(\frac{d\nu_0}{dP_0}\right)\right] + \nu_\alpha(B),$$

while the RHS is written as:

$$E^{P_\alpha}\left[V\left(\frac{d\nu_\alpha}{dP_\alpha}\right)\right]+\nu_\alpha(B).$$

Noting that $\alpha(dP_1/dP_\alpha)+(1-\alpha)(dP_0/dP_\alpha) = 1$ and $\alpha(dP_1/dP_\alpha)(d\nu_1/dP_1)+(1-\alpha)(dP_0/dP_\alpha)(d\nu_0/dP_0)=d\nu_\alpha/dP_\alpha$, the strict convexity of V shows that (4.14) holds only if

$$\frac{d\nu_1}{dP_1}=\frac{d\nu_0}{dP_0},\ \text{a.s. on } A:=\left\{\frac{dP_1}{dP_\alpha}>0 \text{ and } \frac{dP_0}{dP_\alpha}>0\right\}.$$

Since $d\nu_0/dP_0 = d\nu_1/dP_1 = 0$ outside A by $\nu_0\sim\nu_1\sim P_0\sim P_1$ and the definition of the Radon–Nikodym densities, we have (4.13). □

5. Optimal strategy

5.1. Proof of Theorem 2.8 (a)

Let $\theta\in\Theta_{V,\hat{P}}$, i.e., $\theta\cdot S$ is a supermartingale under $\forall Q\in\mathcal{M}_V(\hat{P})$. Then

$$\begin{aligned}\inf_{P\in\mathcal{P}} E^P[U(\theta\cdot S_T+B)] &\le E^{\hat{P}}[U(\theta\cdot S_T+B)]\\ &\le E^{\hat{P}}\left[V\left(\hat{\lambda}\frac{d\hat{Q}}{d\hat{P}}\right)+\hat{\lambda}\frac{d\hat{Q}}{d\hat{P}}(\theta\cdot S_T+B)\right]\\ &\le E^{\hat{P}}\left[V\left(\hat{\lambda}\frac{d\hat{Q}}{d\hat{P}}\right)+\hat{\lambda}\frac{d\hat{Q}}{d\hat{P}}B\right]\\ &= V(\hat{\lambda}\hat{Q}|\hat{P})+\hat{\lambda}E^{\hat{Q}}[B]\\ &= \inf_{\lambda>0}\inf_{(Q,P)\in\mathcal{M}_V\times\mathcal{P}}(V(\lambda Q|P)+\lambda E^Q[B]).\end{aligned}$$

Here the second inequality follows from Young's inequality, while the third from the fact that $\theta\cdot S$ is a $\hat{Q}$-supermartingale.

On the other hand, since $\Theta_{bb}\subset\Theta_{V,\hat{P}}$,

$$\begin{aligned}\sup_{\theta\in\Theta_{V,\hat{P}}}\inf_{P\in\mathcal{P}} E^P[U(\theta\cdot S_T+B)] &\ge \sup_{\theta\in\Theta_{bb}}\inf_{P\in\mathcal{P}} E^P[U(\theta\cdot S_T+B)]\\ &= \inf_{\lambda>0}\inf_{(Q,P)\in\mathcal{M}_V\times\mathcal{P}}(V(\lambda Q|P)+\lambda E^Q[B]),\end{aligned}$$

by Theorem 2.4. Thus we have (2.21).

5.2. Proof of Theorem 2.8 (b), (c)

The supermartingale property of $\hat{\theta} \cdot S$ under all $Q \in \mathcal{M}_V(\hat{P})$ will be verified by considering a dynamic version of variational inequality for $\hat{Q}$, using the *m-stability* (*multiplicative* stability, see Delbaen [7], Definition 1) of the set of local martingale measures. This is more or less a standard route, and similar arguments are found in literature with slight differences in assumptions. We follow the line of Föllmer and Gundel [12], Lemma 3.12, with modifications involving the presence of unbounded endowment B.

Let $(\hat{\nu}, \hat{P}) = (\hat{\lambda}\hat{Q}, \hat{P})$ be a solution to the dual problem, and $\hat{\theta}$ be the integrand appearing the representation (2.19). Note that $\hat{Q}$ is also a minimizer of $Q \mapsto V(\hat{\lambda}Q|\hat{P}) + \hat{\lambda}E^Q[B]$. Let $\hat{Z}_t := (d\hat{Q}/d\hat{P})|_{\mathcal{F}_t}$, and for each Q, we set $Z_t^Q := (dQ/d\hat{P})|_{\mathcal{F}_t}$.

Lemma 5.1. *For all stopping times $\tau \le T$ and $Q \in \mathcal{M}_V^e(\hat{P})$,*

$$E^{\hat{P}}\left[V\left(\hat{\lambda}\hat{Z}_T\right) + \hat{\lambda}\hat{Z}_T B \mid \mathcal{F}_\tau\right] \le E^{\hat{P}}\left[V\left(\hat{\lambda}\hat{Z}_\tau\frac{Z_T^Q}{Z_\tau^Q}\right) + \hat{\lambda}\hat{Z}_\tau\frac{Z_T^Q}{Z_\tau^Q}B \mid \mathcal{F}_\tau\right]. \tag{5.1}$$

Proof. We first show that the RHS of (5.1) is well-defined and finite, $\hat{P}$-a.s. For each n, set

$$A_n := \{\hat{Z}_\tau/Z_\tau^Q \in (1/n, n)\} \in \mathcal{F}_\tau. \tag{5.2}$$

Since $\hat{Q}, Q \sim \hat{P}$, $\hat{Z}_\tau/Z_\tau^Q$ is finite and strictly positive, $\hat{P}$-a.s., hence $\lim_n \hat{P}(A_n) = 1$. By (2.5), there exist constants $K_n, K_n' > 0$, for each n, such that

$$V\left(\hat{\lambda}\frac{\hat{Z}_\tau}{Z_\tau^Q}Z_T^Q\right) \le K_n V(Z_T^Q) + K_n'(Z_T^Q + 1) \in L^1(\hat{P}), \text{ on } A_n. \tag{5.3}$$

Thus $E^{\hat{P}}[V(\hat{\lambda}\hat{Z}_\tau(Z_T^Q/Z_\tau^Q))|\mathcal{F}_\tau] < \infty$ a.s. on each A_n, hence it is finite a.s. On the other hand, $E^{\hat{P}}[\hat{\lambda}\hat{Z}_\tau(Z_T^Q/Z_\tau^Q)|B||\mathcal{F}_\tau] = \hat{\lambda}\hat{Z}_\tau E^Q[|B||\mathcal{F}_\tau] < \infty$ a.s., since $B \in L^1(Q)$.

Set

$$C := \left\{E^{\hat{P}}\left[V\left(\hat{\lambda}\hat{Z}_T\right) + \hat{\lambda}\hat{Z}_T B \mid \mathcal{F}_\tau\right] > E^{\hat{P}}\left[V\left(\hat{\lambda}\hat{Z}_\tau\frac{Z_T^Q}{Z_\tau^Q}\right) + \hat{\lambda}\hat{Z}_\tau\frac{Z_T^Q}{Z_\tau^Q}B \mid \mathcal{F}_\tau\right]\right\}$$

Suppose that $\hat{P}(C) > 0$. Then taking a large n, $\hat{P}(C \cap A_n) > 0$. Setting $A := A_n \cap C$, we define a new probability measure $\bar{Q}$ with density process $\bar{Z}$ by:

$$\bar{Z}_T := 1_{A^c}\hat{Z}_T + 1_A \hat{Z}_\tau \frac{Z_T^Q}{Z_\tau^Q}.$$

Note first that $\mathcal{M}_{loc}(\hat{P})$ is *m-stable* by [7, Proposition 5], hence $\bar{Q} \in \mathcal{M}_{loc}(\hat{P})$. Also, using (5.3), we have $\bar{Q} \in \mathcal{M}_V(\hat{P})$. Finally,

$$\begin{aligned}
&V(\hat{\lambda}\bar{Q}|\hat{P}) + \hat{\lambda}E^{\bar{Q}}[B] \\
&= E^{\hat{P}}\left[1_{A^c}E^{\hat{P}}[V(\hat{\lambda}\hat{Z}_T) + \hat{\lambda}\hat{Z}_T B|\mathcal{F}_\tau]\right] \\
&\quad + E^{\hat{P}}\left[1_A E^{\hat{P}}\left[V\left(\hat{\lambda}\hat{Z}_\tau \frac{Z_T^Q}{Z_\tau^Q}\right) + \hat{\lambda}\hat{Z}_\tau \frac{Z_T^Q}{Z_\tau^Q} B \,\Big|\, \mathcal{F}_\tau\right]\right] \\
&< E^{\hat{P}}\left[1_{A^c}E^{\hat{P}}[V(\hat{\lambda}\hat{Z}_T) + \hat{\lambda}\hat{Z}_T B|\mathcal{F}_\tau]\right] \\
&\quad + E^{\hat{P}}\left[1_A E^{\hat{P}}[V(\hat{\lambda}\hat{Z}_T) + \hat{\lambda}\hat{Z}_T B|\mathcal{F}_\tau]\right] \\
&= V(\hat{\lambda}\hat{Q}|\hat{P}) + \hat{\lambda}E^{\hat{Q}}[B].
\end{aligned}$$

This contradicts to the minimality of $\hat{Q}$, hence $\hat{P}(C) = 0$. □

Lemma 5.2. *For any $Q \in \mathcal{M}_V(\hat{P})$, and $\tau \le T$,*

$$E^{\hat{Q}}[V'(\hat{\lambda}\hat{Z}_T) + B|\mathcal{F}_\tau] \le E^Q[V'(\hat{\lambda}\hat{Z}_T) + B|\mathcal{F}_\tau], \quad Q\text{-a.s.} \tag{5.4}$$

Proof. Since both sides of (5.4) are $\mathcal{F}_\tau$-measurable, it suffices to show that the desired inequality holds $\hat{P}$-a.s. on the set $\{Z_\tau^Q > 0\}$, or equivalently:

$$Z_\tau^Q\left\{E^Q[V'(\hat{\lambda}\hat{Z}_T) + B|\mathcal{F}_\tau] - E^{\hat{Q}}[V'(\hat{\lambda}\hat{Z}_T) + B|\mathcal{F}_\tau]\right\} \ge 0, \quad \hat{P}\text{-a.s.} \tag{5.5}$$

Let $Q \in \mathcal{M}_V(\hat{P})$, $Q_\alpha := \alpha Q + (1-\alpha)\hat{Q}$ for each $\alpha \in [0,1)$, and Z^α the density process of Q_α w.r.t. $\hat{P}$. Note that $Q_\alpha \in \mathcal{M}_V^e(\hat{P})$, since $\hat{Q} \sim \hat{P}$. Therefore, Lemma 5.1 shows: for each $\alpha \in [0,1)$,

$$E^{\hat{P}}[V(\hat{\lambda}\hat{Z}_T) + \hat{\lambda}\hat{Z}_T B|\mathcal{F}_\tau] \le E^{\hat{P}}\left[V\left(\hat{\lambda}\hat{Z}_\tau \frac{Z_T^\alpha}{Z_\tau^\alpha}\right) + \hat{\lambda}\hat{Z}_\tau \frac{Z_T^\alpha}{Z_\tau^\alpha} B \,\Big|\, \mathcal{F}_\tau\right]. \tag{5.6}$$

We first compute Z^α_T/Z^α_τ.

$$\begin{aligned}
\frac{Z^\alpha_T}{Z^\alpha_\tau} &= \frac{\alpha Z^Q_T + (1-\alpha)\hat{Z}_T}{\alpha Z^Q_\tau + (1-\alpha)\hat{Z}_\tau} \\
&= \frac{\alpha Z^Q_\tau}{\alpha Z^Q_\tau + (1-\alpha)\hat{Z}_\tau} \frac{Z^Q_T}{Z^Q_\tau} + \frac{(1-\alpha)\hat{Z}_\tau}{\alpha Z^Q_\tau + (1-\alpha)\hat{Z}_\tau} \frac{\hat{Z}_T}{\hat{Z}_\tau} \\
&= \Lambda(\alpha)\frac{Z^Q_T}{Z^Q_\tau} + (1-\Lambda(\alpha))\frac{\hat{Z}_T}{\hat{Z}_\tau},
\end{aligned}$$

where $\Lambda(\alpha) := \alpha Z^Q_\tau/(\alpha Z^Q_\tau + (1-\alpha)\hat{Z}_\tau)$. $\Lambda(\alpha)$ is $\mathcal{F}_\tau$-measurable for each α, while $\alpha \mapsto \Lambda(\alpha)$ is increasing, differentiable, and

$$\Lambda(0) = 0, \quad \Lambda'(0) = \frac{Z^Q_\tau}{\hat{Z}_\tau}.$$

Set

$$G(\alpha) := V\left(\hat{\lambda}\hat{Z}_\tau \frac{Z^\alpha_T}{Z^\alpha_\tau}\right) + \hat{\lambda}\hat{Z}_\tau \frac{Z^\alpha_T}{Z^\alpha_\tau} B.$$

We see that on $\{Z^Q_\tau > 0\}$,

$$\frac{G(\alpha) - G(0)}{\Lambda(\alpha)} \searrow \hat{\lambda}\hat{Z}_\tau \left\{V'(\hat{\lambda}\hat{Z}_T) + B\right\}\left(\frac{Z^Q_T}{Z^Q_\tau} - \frac{\hat{Z}_T}{\hat{Z}_\tau}\right).$$

Therefore, the conditional monotone convergence theorem together with (5.6) shows:

$$\begin{aligned}
0 &\le E^{\hat{P}}\left[\frac{G(\alpha) - G(0)}{\alpha} \,\Big|\, \mathcal{F}_\tau\right] \\
&= \frac{\Lambda(\alpha)}{\alpha} E^{\hat{P}}\left[\frac{G(\alpha) - G(0)}{\Lambda(\alpha)} \,\Big|\, \mathcal{F}_\tau\right] \\
&\to \frac{Z^Q_\tau}{\hat{Z}_\tau} E^{\hat{P}}\left[\hat{\lambda}\hat{Z}_\tau \left\{V'(\hat{\lambda}\hat{Z}_T) + B\right\}\left(\frac{Z^Q_T}{Z^Q_\tau} - \frac{\hat{Z}_T}{\hat{Z}_\tau}\right) \,\Big|\, \mathcal{F}_\tau\right] \\
&= \hat{\lambda} Z^Q_\tau E^{\hat{P}}\left[\left\{V'(\hat{\lambda}\hat{Z}_T) + B\right\}\left(\frac{Z^Q_T}{Z^Q_\tau} - \frac{\hat{Z}_T}{\hat{Z}_\tau}\right) \,\Big|\, \mathcal{F}_\tau\right],
\end{aligned}$$

and we have (5.5) by the Bayes' formula. □

Remark 5.3. The inequality (5.4) can not be extended beyond the support of Q. In fact, $G(\alpha) \equiv V(\hat{\lambda}\hat{Z}_T) + \hat{\lambda}\hat{Z}_T B$ on $\{Z^Q_\tau = 0\}$, hence (5.6) tells us nothing outside the support of Q.

Proof of Theorem 2.8 (b). We know from Theorem 2.7 (b) that $\hat{\theta} \cdot S$ is a $\hat{Q}$-martingale and

$$V'(\hat{\lambda} d\hat{Q}/d\hat{P}) = -(\hat{\theta} \cdot S_T + B) \quad \Leftrightarrow \quad V'(\hat{\lambda} d\hat{Q}/d\hat{P}) + B = -\hat{\theta} \cdot S_T, \quad \hat{P}\text{-a.s.}$$

Thus, for each $Q \in \mathcal{M}_V(\hat{P})$ and $\tau \leq T$, Lemma 5.2 shows:

$$\begin{aligned} \hat{\theta} \cdot S_\tau &= -E^{\hat{Q}}[V'(\hat{\lambda} d\hat{Q}/d\hat{P}) + B|\mathcal{F}_\tau] \\ &\geq -E^{Q}[V'(\hat{\lambda} d\hat{Q}/d\hat{P}) + B|\mathcal{F}_\tau], \quad Q\text{-a.s.}, \end{aligned}$$

i.e., the stochastic integral $\hat{\theta} \cdot S$ is bounded from below by a Q-martingale. We have therefore $\hat{\theta} \cdot S$ is a Q-supermartingale as a local martingale (by [2]) bounded from below by a martingale. □

Proof of Theorem 2.8 (c). By the additional assumption $\hat{Q} \sim \hat{P} \sim \mathbb{P}$, and part (b), we have $\hat{\theta} \in \Theta_{V,\hat{P}}$, and the representation (2.19) holds $\mathbb{P}$-a.s., hence P-a.s. for all $P \in \mathcal{P}$. Also, by (A1′) and Lemma 3.3, the variational inequality (4.8) applies to all $P \in \mathcal{P}$ without the *bizarre term* $V(0)P(d\hat{P}/d\mathbb{P} = 0)$.

Now using the relation $U(-V'(y)) = V(y) - yV'(y)$ for all $y > 0$, which follows from the definition of V and (A1′), the inequality (4.8) is rewritten as:

$$E^{\hat{P}}[U(\hat{\theta} \cdot S_T + B)] = \min_{P \in \mathcal{P}} E^{P}[U(\hat{\theta} \cdot S_T + B)] \tag{5.7}$$

Finally, we have

$$\begin{aligned} E^{\hat{P}}[U(\hat{\theta} \cdot S_T + B)] &= E^{\hat{P}}\left[V\left(\hat{\lambda}\frac{d\hat{Q}}{d\hat{P}}\right) - \hat{\lambda}\frac{d\hat{Q}}{d\hat{P}} V'\left(\hat{\lambda}\frac{d\hat{Q}}{d\hat{P}}\right)\right] \\ &= E^{\hat{P}}\left[V\left(\hat{\lambda}\frac{d\hat{Q}}{d\hat{P}}\right) + \hat{\lambda}\frac{d\hat{Q}}{d\hat{P}}(\hat{\theta} \cdot S_T + B)\right] \\ &= V(\hat{\lambda}\hat{Q}|\hat{P}) + \hat{\lambda} E^{\hat{Q}}[B] \\ &= \sup_{\theta \in \Theta_{V,\hat{P}}} E^{\hat{P}}[U(\theta \cdot S_T + B)]. \end{aligned}$$

Here the last equality follows from the computation in the beginning of the proof of Theorem 2.8 (a). Therefore, $(\hat{\theta}, \hat{P})$ is a saddle point. □

References

1. Aliprantis, C.D., Border, K.C.: Infinite Dimensional Analysis. A Hitchhiker's Guide, 3rd edn. Springer, Berlin (2006)

2. Ansel, J.-P., Stricker, C.: Couverture des actifs contingents et prix maximum. Ann. Inst. Henri Poincaré **30**, 303–315 (1994)
3. Becherer, D.: Rational hedging and valuation of integrated risks under constant absolute risk aversion. Insur. Math. Econ. **33**, 1–28 (2003)
4. Bellini, F., Frittelli, M.: On the existence of minimax martingale measures. Math. Finance **12**, 1–21 (2002)
5. Biagini, S., Frittelli, M.: On the extension of the Namioka–Klee theorem and on the Fatou property for risk measures. In: Delbaen, F., Rasonyi, M., Stricker, C. (eds.) Optimality and Risk: Modern Trends in Mathematical Finance. The Kabanov Festschrift, pp. 1–29. Springer, Berlin (2009)
6. Biagini, S., Frittelli, M., Grasselli, M.: Indifference price with general semimartingales. Math. Finance (2010, forthcoming)
7. Delbaen, F.: The structure of m-stable sets in particular of the set of risk neutral measures. In: In Memoriam Paul-Andé Meyer, Séminaire de Probabilités XXXIX. Lecture Notes in Mathematics, vol. 1874, pp. 215–258. Springer, Berlin (2006)
8. Delbaen, F., Schachermayer, W.: A general version of the fundamental theorem of asset pricing. Math. Ann. **300**, 463–520 (1994)
9. Delbaen, F., Schachermayer, W.: A compactness principle for bounded sequences of martingales with applications. In: Seminar on Stochastic Analysis, Random Fields and Applications (Ascona, 1996). Progr. Probab., vol. 45, pp. 137–173. Birkhäuser, Basel (1999)
10. Delbaen, F., Grandits, P., Rheinländer, T., Samperi, D., Schweizer, M., Stricker, C.: Exponential hedging and entropic penalties. Math. Finance **12**, 99–123 (2002)
11. Dellacherie, C., Meyer, P.-A.: Probabilities and Potential A. North-Holland Math. Stud., vol. 29. North-Holland, Amsterdam (1978)
12. Föllmer, H., Gundel, A.: Robust projections in the class of martingale measures. Illinois J. Math. **50**, 439–472 (2006)
13. Föllmer, H., Schied, A., Weber, S.: Robust preferences and robust portfolio choice. In: Bensoussan, A., Zhang, Q., Ciarlet, P.G. (eds.) Mathematical Modelling and Numerical Methods in Finance. Handbook of Numerical Analysis, vol. 15, pp. 29–88. North-Holland, Amsterdam (2009)
14. Frittelli, M., Rosazza Gianin, E.: Equivalent formulations of reasonable asymptotic elasticity. Tech. Rep. 12, Dept. Matematica per le Decisioni, University of Florence (2004)
15. Goll, T., Rüschendorf, L.: Minimax and minimal distance martingale measures and their relationship to portfolio optimization. Finance Stochast. **5**, 557–581 (2001)
16. Hernández-Hernández, D., Schied, A.: Robust utility maximization in a stochastic factor model. Stat. Decis. **24**, 109–125 (2006)

17. Hernández-Hernández, D., Schied, A.: A control approach to robust utility maximization with logarithmic utility and time-consistent penalties. Stoch. Process. Appl. **117**, 980–1000 (2007)
18. Hodges, S.D., Neuberger, A.: Optimal replication of contingent claims under transaction costs. Rev. Futures Markets **8**, 222–239 (1989)
19. Jacod, J.: Calcul stochastique et problèmes de martingales. Lecture Notes in Mathematics, vol. 714. Springer, Berlin (1979)
20. Jacod, J.: Intégrales stochastiques par rapport à une semi-martingale vectorielle et changements de filtration. In: Séminaire de Probabilités XIV. Lecture Notes in Mathematics, vol. 784, pp. 161–172. Springer, Berlin (1980)
21. Kabanov, Y.M., Stricker, C.: On the optimal portfolio for the exponential utility maximization: remarks to the six-author paper. Math. Finance **12**, 125–134 (2002)
22. Kramkov, D., Schachermayer, W.: The asymptotic elasticity of utility functions and optimal investment in incomplete markets. Ann. Appl. Probab. **9**, 904–950 (1999)
23. Kramkov, D., Schachermayer, W.: Necessary and sufficient conditions in the problem of optimal investment in incomplete markets. Ann. Appl. Probab. **13**, 1504–1516 (2003)
24. Mania, M., Schweizer, M.: Dynamic exponential utility indifference valuation. Ann. Appl. Probab. **15**, 2113–2143 (2005)
25. Müller, M.: Market Completion and Robust Utility Maximization. Ph.D. thesis, Humboldt Universität zu Berlin (2005)
26. Owari, K.: Robust exponential hedging in a Brownian setting. JSIAM Lett. **1**, 64–67 (2009)
27. Owari, K.: A note on utility maximization with unbounded random endowment. Asia-Pacific Financial Markets. doi: 10.1007/s10690-010-9122-4 (2010, forthcoming)
28. Owari, K.: Robust exponential hedging and indifference valuation. Int. J. Theor. Appl. Finance (2010, forthcoming)
29. Owen, M.P., Žitković, G.: Optimal investment with an unbounded random endowment and utility-based pricing. Math. Finance **19**, 129–159 (2009)
30. Quenez, M.-C.: Optimal portfolio in a multiple-priors model. In: Dalang, R.C., Dozzi, M., Russo, F. (eds.) Seminar on Stochastic Analysis, Random Fields and Applications IV. Progr. Probab., vol. 58, pp. 291–321. Birkhäuser, Basel (2004)
31. Rouge, R., El Karoui, N.: Pricing via utility maximization and entropy. Math. Finance **10**, 259–276 (2000)
32. Schachermayer, W.: Optimal investment in incomplete markets when wealth may become negative. Ann. Appl. Probab. **11**, 694–734 (2001)

33. Schachermayer, W.: A super-martingale property of the optimal portfolio process. Finance Stochast. **7**, 433–456 (2003)
34. Schied, A.: Optimal investment for risk- and ambiguity-averse preferences: a duality approach. Finance Stochast. **11**, 107–129 (2007)
35. Schied, A., Wu, C.-T.: Duality theory for optimal investment under model uncertainty. Stat. Decis. **23**, 199–217 (2005)
36. Simons, S.: From Hahn–Banach to monotonicity, 2nd edn. Lecture Notes in Mathematics, vol. 1693. Springer, Berlin (2008)
37. Wittmüss, W.: Robust optimization of cousumption with random endowment. Stochastics **80**, 459–475 (2008)
38. Yor, M.: Sous-espaces denses dans L^1 ou H^1 et representation des martingales. In: Séminaire de Probabilités XII. Lecture Notes in Mathematics, vol. 649, pp. 265–309. Springer, Berlin (1978)

Adv. Math. Econ. 14, 183–196 (2011)

Advances in
MATHEMATICAL ECONOMICS

A time-embedded approach to economic equilibrium with incomplete financial markets

A. Jofré[1], R.T. Rockafellar[2], and R.J.-B. Wets[3]

[1] Center for Mathematical Modelling, Department of Mathematical Engineering, University of Chile, Casilla 170/3, Correo 3, Santiago, Chile
(e-mail: ajofre@dim.uchile.cl)

[2] Department of Mathematics, University of Washington, Seattle, WA 98195-4350, USA
(e-mail: rtr@math.washington.edu)

[3] Department of Mathematics, University of California, Davis, CA 95616, USA
(e-mail: rjbwets@ucdavis.edu)

Received: April 27, 2010
Revised: July 13, 2010

JEL classification: D53

Mathematics Subject Classification (2010): 91B50

Abstract. In the models of multi-stage equilibrium with uncertain financial markets that have so far been formulated in extension of the classical Walrasian model with only a single stage, each state is completely isolated in its activity. If there is production, it ends in the state in which it begins. Goods that are not consumed within a state merely perish. Nothing can carry over from one period to the next like money might, and comparisons between the units of account in different states may be problematical.

This paper furnishes a two-stage model in which money assists as a special good which agents like to retain instead of consume. Holding money influences utility through pleasures of wealth and in safeguarding against completely unforseen events. The agents' planning is able in this way also to reflect continuing expectations of the future beyond the second stage.

The ability of agents to retain money furthermore has the technical benefit that the survivability conditions needed to establish the existence of equilibrium can reduced to a very simple and yet much more appealing form than has been discerned until now. Endogenous transaction costs on the sellers of financial contracts help in this as well.

Key words: general economic equilibrium, incomplete financial markets, time embedding, retention of money, transaction costs, ample survivability

DOI: 10.1007/978-4-431-53883-7_8,

1. Introduction

Many conceptual hurdles must be faced in the modeling of economic equilibrium. Everyone understands that the very idea of equilibrium is an artificial abstraction in the ongoing world of economic activity. Nonetheless, this idea has the potential to help greatly in identifying sources of stability and instability along with the main features of market behavior and the conditions on which they depend.

The basic goal in formulating and analyzing a mathematical model of equilibrium is to determine the extent to which *market prices* exist which will *balance supplies with demands* when the agents in an economy *optimize their activities according to their preferences*. However, models can differ in the kinds of markets and the ranges of activities that they try to encompass, and in their degree of success in capturing economic realities.

The celebrated Walrasian model of equilibrium, placed on a modern mathematical footing by Arrow and Debreu [1], has yielded valuable insights but is very limited in its scope. It is *isolated in time* without any past or future, and is therefore unable deal with the large component of economic behavior concerned with making arrangements beyond the needs and incentives of the immediate present. However, any extension to future planning must confront *uncertainty*, and that raises another set of touchy issues.

Financial markets can assist by furnishing agents with possibilities to buy and sell contracts in the present which affect their circumstances in the future. Exchanges of value between different time periods are then enabled, but it is not realistic to expect such markets to be able to hedge or ensure precisely against *all* conceivable uncertainties. Thus, financial markets must be modeled as *incomplete*. That leads to mathematical obstacles which preclude the straightforward approach of Arrow and Debreu. Moreover, some of the central conclusions drawn from the classical one-stage model, such as the capability of market prices to achieve Pareto optimality in the distribution of resources, are then called into question.

One of the daunting challenges in financial markets is that the promises incorporated in the contracts to be bought and sold need to *enforced* somehow when they come due. But that entails, in the first place, being sure that the promises are realistic. Future supplies and demands must be taken into account, yet those supplies and demands will depend on the present and future actions of the various agents. How then can there be any knowledge of that on which an individual agent might plan? Indeed, how can prices for the contracts in a financial market be brought into line without some grasp of future market prices for goods, which in turn ought to reflect balances in future supply and demand?

The availability of information about the future is thus a critical consideration. But it cannot be supposed that the agents have a common view even of likelihoods. Indeed, no representation of the future can ever deserve full trust. Too much will inevitably be unforseen. Moreover, this deeper level uncertainty is bound to affect attitudes of the agents in a Keynesian manner. As explained by Skidelsky [19, Chap. 4]: "for Keynes, money was a 'store of value' as well as a means of transactions; it was 'above all a subtle device for linking the present and the future'." The model presented here tries for the first time to account for this in the modeling of equilibrium by allowing utility of an agent to benefit from the retention of money. By treating money as a special kind of good which can store value, it skirts the customary view of the worth of goods being tied solely to their consumption.

2. Background

To set the stage for the new developments that are the focus of this paper, we first recall the one-stage Arrow–Debreu model of economic equilibrium. We concentrate on a basic version in which agents can exchange goods but firms and production activities are left out. In two-stage models with uncertainty, such as we will come to later, the modeling of intertemporal production carried out by "firms" is an especially difficult matter. Progress is being made, cf. [3], but production will not be considered in this paper.

Agents, indexed by $i = 1, \dots, m$, will exchange goods indexed by $l = 1, \dots, L$ at market prices $p_l \geq 0$. Agent i starts with a goods vector $x_i^0 \in I\!R_+^L$ with components x_{il}^0 and trades it for a goods vector $x_i \in I\!R_+^L$ with components x_{il} subject to the budget constraint $p{\cdot}x_i \leq p{\cdot}x_i^0$, where $p = (p_1, \dots, p_L)$. In doing so, agent i maximizes the utility $u_i(x_i)$ of x_i. The function u_i is taken here to be defined on all of $I\!R^L$ with values in $[-\infty, \infty)$ and nonempty effective domain

$$U_i = \operatorname{dom} u_i = \{\, x_i \mid u_i(x_i) > -\infty \,\} \subset I\!R_+^L.$$

The constraint $x_i \in U_i$ is implicit then in the maximization; we speak of U_i as the *survival* set for agent i. It is assumed that u_i is nondecreasing and upper semicontinuous (usc); that requires sets of the form $\{\, x_i \mid u_i(x_i) \geq \alpha \,\}$ to be closed for all $\alpha \in I\!R$, but does not necessitate U_i itself being closed.

Definition: classical equilibrium. *A equilibrium in this setting of pure exchange is comprised of a price vector $\bar{p}$ and goods vectors $\bar{x}_i$ for $i = 1, \dots, m$ such that*

(a) *$\bar{x}_i$ maximizes u_i under the budget dictated by $\bar{p}$.*

(b) *The markets clear:* $\sum_{i=1}^m x_i^0 - \sum_{i=1}^m \bar{x}_i \geq 0,\ \bar{p}{\cdot}\Big[\sum_{i=1}^m x_i^0 - \sum_{i-1}^m \bar{x}_i\Big] = 0.$

The market clearing conditions are written in a convenient vector form. They reflect free disposal and amount to requiring for each good l that $\sum_{i=1}^{m} x_{il}^{0} = \sum_{i=1}^{m} \bar{x}_{il}$ unless $\bar{p}_l = 0$, in which case $\sum_{i=1}^{m} x_{il}^{0} \geq \sum_{i=1}^{m} \bar{x}_{il}$ is allowed.

An issue that comes up in connection with establishing the existence of such an equilibrium is the extent to which the agents can, without trading anything, at least *survive*. Plain survival only refers to having $x_i^0 \in U_i$, but existence arguments generally require more than just that. The following standard result, in which U_i is replaced by its interior, $\operatorname{int} U_i$, in the survivability condition, provides a basis for comparisons. It also draws on the concept of *insatiability* of utility in the sense of there being no goods vector at which u_i attains a maximum value over U_i.

Theorem of Arrow and Debreu [1]. *For utility functions u_i that are insatiable, quasi-concave and continuous relative to the sets U_i, a classical equilibrium exists under the strong survivability assumption that*

$$x_i^0 \in \operatorname{int} U_i \quad \textit{for all agents } i.$$

Although strong survivability is a simple condition that suffices to obtain the existence of equilibrium, it is distressingly restrictive. Insistence on having $x_i^0 \in \operatorname{int} U_i \subset I\!R_+^L$ entails that *each agent must start with a positive quantity of every good.* Some technical ways around this were developed in [1] and elaborated later for instance by Florig [7, 8], but they are quite complicated and their general economic implications are not easy to fathom.

The mathematical reason why strong survivability comes into play is usually understood by economists in the context of properties needed in fixed-point arguments, but it is equally understandable from the perspective of optimization theory. In problems of optimization with constraints, such as the one faced by agent *i*, some "constraint qualification" is typically required. In this setting of convexity, that would most naturally be the *Slater condition*: there should be an $\hat{x}_i \in U_i$ such that $p{\cdot}x_i < p{\cdot}x_i^0$. Then too a Lagrange multiplier for the budget constraint will be available.

The trouble with this constraint qualification, though, is that the price vector p is not fixed in advance and will only be determined by the equilibrium. The Slater condition might seemingly have to be guaranteed for *every* possible price vector $p \geq 0$, $p \neq 0$. The only way to do that is through strong survivability.

If some of the price components of p may be 0, and agent i happens only to start with goods that are in that case worthless, then the Slater condition would definitely fail. But even with positive prices, it could fail if survivability can only effected with the budget constraint tight.

Questions of about the prices that may prevail in equilibrium are significant from other angles as well. If the price of a particular good comes out as 0, does that mean no agent has any desire for it, or at least no marginal desire for more of it in the equilibrium state? The answer is yes on the basis of Lagrange multiplier analysis of optimality in an agent's utility problem, but that leads back to the need for a constraint qualification supporting the existence of a Lagrange multiplier for the budget constraint. In that case anyway, the positivity of a good's price can be ensured by an assumption on the utility function that prevents marginal utility of that good from ever being 0. This would only have to be true for at least one agent i. Of course, it is desirable to keep in a mode where an agent may only be attracted to a subset of the goods.

Marginal utility may seem to involve derivatives, but they might just be one-sided directional derivatives, which would be available without any differentiability assumption if the functions u_i are concave.

Another feature of the classical model is that the price vector p only provides *relative* prices, which are unscaled, instead of *money* prices. If the price of a particular good is positive in equilibrium, then that good can serve as a *numéraire* having unit price scaled to 1. However, that does not lead to money prices unless "money" can be viewed somehow as a "good." Indeed, money generally fits the picture of being limited in supply and freely tradable in markets, but if included as a "good," would it be able to serve as a numéraire? Not unless its price relative to other goods turned out to be positive – that returns us to the preceding discussion of utility.

The fundamental difficulty is that the classical framework ties utility to consumption *only*, and the marginal utility of "consuming money" in competition with other consuming goods is doubtful, although perhaps salvageable through some interpretation. *We see this largely as an artifact of the peculiar nature of a one-period model of equilibrium without past or future, since much of the economic importance of money revolves around its role in connecting past and future with the present.*

In what follows, we present a time-embedded two-period model of equilibrium which does give value to money. It is a realization of a broader model we have put together in [12] with many additional features. By getting to the heart of the main issues, we hope it will help in making the innovations clearer.

3. Incorporating an uncertain future

As a general principle, the future is uncertain to agents making decisions in the present, and any two-stage model of equilibrium must reflect that. Agents need to plan for the future despite its uncertainty. They ought to be able to do that through buying and selling *contracts* that promise deliveries or exchanges of goods in the future. However, for that to make sense there must be some way of assessing the present value of such contracts.

Especially, agents should be able to *borrow and lend money* in the present, settling accounts in the future. The interest rates in that operation ought to relate to future purchasing power. On the other hand, the future prices of goods ought to relate to future supply and demand. Thus, future markets in goods inevitably need to be contemplated.

But how should information about the future be generated for making decisions in the present? Can anything be learned from a model with just one stage of future? No matter how many stages, modeling issues will arise over having a fixed time horizon. Our idea is to mitigate the difficulties over this by an expanded view of an agent's wishes and actions. In fact, we see the incorporation of multiple future stages as perhaps detrimental to rather than beneficial to reality by overloading the structure of decisions and undermining the plausibility of the information behind them.

Although contracts for the delivery of all kinds of goods would be good to include in the model, we restrict ourselves here to money alone. That will make the pattern of our development easier to understand. A broader model with additional features is given in our paper [12].

The model here has states $s = 0, 1, \dots, S$, with $s = 0$ standing for the present (at time 0) and $s = 1, \dots, S$ standing for the different futures being taken into consideration (at time 1) because of uncertainty. The agents can buy and sell goods in all of these states, but they also have money and can pass it from the present to the future. Money is not "consumed"; instead, agents like to *retain* it.

Specifically, agent i, in each state s, consumes a goods vector $x_i(s) \geq 0$ and retains a money amount $m_i(s)$ in the environment of benefiting from incoming endowments $x_i^0(s) \geq 0$ and $m_i^0(s) \geq 0$. The agent's utility function has the form $u_i(x_i, m_i)$ for

$$x_i = (x_i(0), x_i(1), \dots, x_i(S)) \in (\mathbb{R}_+^L)^{1+S},$$
$$m_i = (m_i(0), m_i(1), \dots, m_i(S)) \in \mathbb{R}_+^{1+S}.$$

We allow u_i to take on $-\infty$ and in that way focus implicitly on a survival set

$$U_i = \{ (x_i, m_i) \mid u_i(x_i, m_i) > -\infty \} \subset (\mathbb{R}_+^L)^{1+S} \times \mathbb{R}_+^{1+S}.$$

We assume that

> u_i is upper semicontinuous, concave and nondecreasing, moreover continuous relative to the sets $\{ (x_i, m_i) \mid u_i(x_i, m_i) \geq c \}$ for $c \in \mathbb{R}$.

Those level sets are closed because of the upper semicontinuity, but U_i might not be closed. (Utility might tend to $-\infty$ as the boundary of U_i is approached.) Having u_i be nondecreasing means that $u_i(x_i, m_i) \leq u_i(x_i', m_i')$

when $(x_i, m_i) \le (x_i', m_i')$. The concavity, in contrast to quasi-concavity, is a slight retreat from the classical setting but it simplifies and enhances the model in many respects. In particular it assists us in working with the following key assumption:

$u_i(x_i, m_i)$ increases on U_i with respect to each of the components $m_i(s)$.

In other words, retaining money is always attractive to every agent i, although it must nevertheless compete with consumption as represented by the components of the goods vectors $x_i(s)$.

The various goods $l = 1, \ldots, L$ can be bought and sold in markets in every state s. These markets are governed by a *money-denominated* price system

$$p = (p(0), p(1), \ldots, p(S)),$$
$$\text{where } p(s) = (p_1(s), \ldots, p_L(s)) \text{ gives the prices in state } s.$$

In state $s = 0$ there is also a financial market in which the agents can participate. The financial market revolves here around contracts in money only, as already mentioned. Such contracts are available in patterns $k = 1, \ldots, K$. A unit of contract k costs q_k in the present and pays $d_k(s)$ in future state s. The payout vectors

$$d_k = (d_k(1), \ldots, d_k(S)), \quad \text{assumed} \neq (0, \ldots, 0),$$

are known data, but the price vector q must be determined together with the price vectors $p(s)$ in the present and future markets of goods. Examples will be provided shortly.

Agent i buys a (generally fractional) amount $z_{ik}^+ \ge 0$ of contract k and sells an amount $z_{ik}^- \ge 0$, thereby putting together a *portfolio*

$$(z_i^+, z_i^-) \text{ with } z_i^+ = (z_{i1}^+, \ldots, z_{ik}^+) \in I\!R_+^K, \quad z_i^- = (z_{i1}^-, \ldots, z_{ik}^-) \in I\!R_+^K.$$

Simultaneously buying and selling a contract k is not prohibited but will be eliminated in optimality by a transaction cost. We introduce this *endogenously* by supposing that the *selling* of a unit of contract k uses up a goods vector $g_k \in I\!R_+^L$ which will end up costing costing $p(0)\cdot g_k$. We assume that

$$g_k \text{ has a component} > 0 \text{ in some good that is}$$
$$\textit{universally attractive} \text{ in the present,}$$

by which we mean that every agent's utility u_i increases on U_i with respect to that component of $x_i(0)$. Let

$$d(s) = (d_1(s), \ldots, d_K(s)), \qquad G = \text{the matrix with columns } g_k.$$

The portfolio (z_i^+, z_i^-) of agent i

costs $q \cdot [z_i^+ - z_i^-] + p(0) \cdot G z_i^-$ in the present, and then

pays $d(s) \cdot [z_i^+ - z_i^-]$ in the future states $s = 1, \ldots, S$.

Precedent for such exogenous transaction costs can be seen in work of Arrow and Hahn [2] and especially Laitenberger [13], although articulated differently. Here, of course, our attention is limited to contracts that deliver money, but the costs enter more broadly in our paper [12] where the contracts can involve delivery of other goods as well.

The optimization problem for agent i takes the form of choosing choosing x_i, m_i and (z_i^+, z_i^-) subject to the budget constraints

$$\begin{aligned} &p(0) \cdot x_i(0) + m_i(0) + q \cdot [z_i^+ - z_i^+] + p(0) \cdot G z_i^- \leq p(0) \cdot x_i^0 + m_i^0(0), \\ &p(s) \cdot x_i(s) + m_i(s) \leq p(0) \cdot x_i^0(s) + m_i^0(s) + d(s) \cdot [z_i^+ - z_i^-] \\ &\quad + m_i(0) \text{ for } s > 0. \end{aligned} \tag{1}$$

Note the term $m_i(0)$ in the budget for the future states s. This conforms to the provision that money retained at time 0 is available with certainty at time 1.

Here are two elementary examples of contracts in our picture. For concreteness of language, they are explained with dollars as money. The transaction cost will be expressed by

$$\delta_k(p) = p(0) \cdot q.$$

Example 1: simple lending and borrowing. *As a particular case, contract k could have*

$$d_k(s) = 1 \text{ for all future states } s = 1, \ldots, S.$$

Buying a unit refers then to giving out q_k dollars in the present and always getting back 1 dollar in the future, with

$$\text{interest rate} = \frac{1}{q_k} - 1.$$

Selling refers instead to getting $q_k - \delta_k(p)$ dollars in the present in return for promising to pay back 1 dollar in the future, with

$$\text{interest rate} = \frac{1}{q_k - \delta_k(p)} - 1.$$

The transaction cost thus induces a natural difference in interest rates between lending and borrowing.

Example 2: simple insurance. *As a particular case, contract k could depend on a specific future state $\bar{s}$ and have*

$$d_k(\bar{s}) = 1, \quad \text{but } d_k(s) = 0 \text{ for all future states } s \neq \bar{s}.$$

Buying a unit of contract k refers then to giving out q_k dollars in the present but getting back 1 dollar only if the future state turns out to be $\bar{s}$, otherwise nothing. Selling refers to acquiring $q_k - \delta_k(p)$ dollars in the present, and only having to pay out 1 dollar if $\bar{s}$ occurs, otherwise nothing..

If simple insurance as in Example 2 is not in the market as one of the contracts k, there it the possibility that it can anyway be replicated by some linear combination of other contracts that are in the market. If this is true for *every* future state $\bar{s}$, the financial market is said to be *complete*. But in reality, financial markets are never complete. This observation has heavily influenced all extensions of equilibrium theory to financial markets after the early proposals of Arrow and Debreu, which did suppose completeness.

4. Existence of equilibrium

The main result about equilibrium in our model with its additional features about money will now be formulated.

Definition: money-supported equilibrium. *This is comprised of elements $\bar{x}_i, \bar{m}_i, (\bar{z}_i^+, \bar{z}_i^-)$, for the agents $i = 1, \ldots, m$ along with price vectors $\bar{p}$ and $\bar{q}$ such that*

(a) *$(\bar{x}_i, \bar{m}_i)$ maximizes $u_i(x_i, m_i)$ under the budget constraints (1) coming from $\bar{p}$ and $\bar{q}$.*
(b) *The goods markets clear in all states $s = 0, 1, \ldots, S$:*

$$\textstyle\sum_{i=1}^m x_i^0(s) - \sum_{i=1}^m \bar{x}_i(s) \geq 0, \quad \bar{p}(s) \cdot \Big[\sum_{i=1}^m x_i^0(s) - \sum_{i=1}^m \bar{x}_i(s) \Big] = 0.$$

(c) *The financial markets clear: $\sum_{i=1}^m \bar{z}_i^+ = \sum_{i=1}^m \bar{z}_i^-$.*
(d) *The money supply is respected and money is conserved:*

$$\sum_{i=1}^m \bar{m}_i(0) = \sum_{i=1}^m m_i^0(0) \text{ for } s = 0,$$

$$\sum_{i=1}^m \bar{m}_i(s) = \sum_{i=1}^m [m_i^0(s) + m_i(0)] \text{ for } s > 0.$$

A virtue of such a money-supported equilibrium is that existence can be ensured without resorting to strong survivability.

Assumption: ample survivability. *The agents i could, if they wished, choose elements $(\hat{x}_i, \hat{m}_i) \in U_i$ such that*

$$\begin{aligned} \hat{x}_i(0) \le x_i^0(0) \;\; &\textit{but} \;\; \hat{m}_i(0) < m_i^0(0) \;\; \textit{for} \;\; s = 0, \\ \hat{x}_i(s) \le x_i^0(s) \;\; &\textit{and} \;\; \hat{m}_i(s) \le m_i^0(s) \;\; \textit{for} \;\; s = 1, \ldots, S, \\ \sum\nolimits_{i=1}^{m} \hat{x}_i(s) < \sum\nolimits_{i=1}^{m} x_i^0(s) \;\; &\textit{for} \;\; s = 0, 1, \ldots, S. \end{aligned} \tag{2}$$

The interpretation of this condition is that the agents would be able to survive, outside of any participation in the markets, *individually* without using all their money in the present, and *collectively* with a surplus remaining in every good. Thus, in contrast to the situation under strong survivability, there is no need for every agent to possess some of every good initially. Nevertheless, a way is opened up for invoking the Slater condition for an agent's budget constraints. This is evident from (2) for the initial budget, but it also follows then effectively for the subsequent budgets because the agent can save some of the initial surplus of money to furnish a buffer in the future.

Existence Theorem. *Under ample survivability and the assumptions on the financial contracts and the agent's utility functions, a money-supported equilibrium exists.*

Insatiability of utility is present here in the assumption that agents are always attracted to retaining money. That attraction is deemed to have a cultural origin based on a history of how money is valued in society and persists in usefulness because of its convenience and liquidity.

We proceed to discuss this new result, which stands as a specialization of the one in our paper [12]. The nature of the markets and how they are viewed is a key issue.

The markets here are modeled, according to custom in the economic literature, as brought into line by *Walrasian brokers* who match supply with demand. This avoids having to view transactions as one-on-one and goes back to Walras himself. Radner [16] in 1972 initiated having separate markets in the present and the future. His agents could promise delivery of goods (in so-called *real* contracts), but not "money" (unless that referred only to a numéraire good). An arbitrary number of uncertain future stages were envisioned, not just one, as here. Artificial (exogenous) bounds on contract sales were needed in order to establish the existence of an equilibrium.

In an influential development in 1985, Cass [4] and Werner [20] independently obtained the existence of an equilibrium in a model in which prices and contracts were all in terms of "money" (so-called *nominal* contracts), but they did not include bounds on money supply. As noted by Magill and Shafer [15], that introduced indeterminancy in which there was no way to relate the money in any one state with the money in another. A proposal of Magill and Quinzii [14] to remedy this required putting the money into a multi-layered scheme of transactions quite different from it behaving as a good like here.

In the models of Radner, Cass and Werner, even though they incorporate an uncertain and potentially multi-stage future, each state s is still isolated, just as in the Walrasian formulation that was the focus of Arrow and Debreu. All the goods that become available in a state s, through endowments or, in some versions, production, must be consumed in that same state. There is no carry-over of goods whatsoever, not to speak of there being money that might be saved.

An important question concerns information. How are the agents supposed to know, at time 0, what will happen in the markets at a later time? The models implicitly require some such knowledge, because *it is in the initial state* $s = 0$ *that the agents already have to plan their buying and selling of goods in the future states* $s > 0$, and their budgets for such planning depend on the prices of goods in those states as well on the prices in the financial market.

An alternative model of *temporary equilibrium*, as explained by Grandmont [9], sets aside decisions about trading goods in future markets. Agents, in their financial planning, act only on "anticipations" of future prices. This is an interesting proposal, but in getting down to serious mathematical details, such as the specifics of "anticipations," it appears to run into thorny difficulties. Fundamentally there is a lack of feedback about future supplies and demands of goods and therefore no way for agents, through some interchange, to tune their "anticipations" to an approximation of reality.

The point we believe should be kept in mind is that the future markets in the models of Radner, Cass and Werner, as well as here, cannot truly be claimed to take place in the future. This is an inevitable consequence of the notion of a Walrasian broker for those markets. That broker is entrusted with matching future supply and demand for goods on the basis of plans for buying and selling goods that the agents fix *in the present*. This broker operation must therefore be in the present as well, not in the future! We see it as a model of information exchange in which an agent can gain insights into what other agents may be hoping to do. That exchange generates feedback about prices, allowing the agents to get a better handle on how to plan for the future.

As a matter of fact, in our view there is no need even to think that, when the real future arrives, the agents will definitely carry out these plans. Circumstances may have changed beyond those taken into account when the plans were made. Our interpretation of equilibrium modeling with an uncertain future is thus a compromise between the traditional "sequential market" approach and that of "temporary equilibrium." However it brings skepticism about the virtues of incorporating more than just a few stages of the future.

That would require a further proliferation of Walrasian brokers trying to glean planning information in the present, which would stretch the information-gathering concept too far.

5. Pattern of development

The model laid out here fits mathematically as a special case of a much more extensive model in our paper [12], and the existence theorem here follows essentially as a corollary of the existence theorem there. Although for that reason the details of its proof are superfluous for the present paper, an indication of the main ideas and their articulation in a nontraditional context may be illuminating.

The strategy is to characterize the existence of equilibrium as a so-called *variational inequality* problem. Such formulations have gained in importance areas of application related to optimization, as in the book of Facchinei and Pang [6]. They rely on many advances in variational analysis, available for example in [18], as an enlargement of convex analysis [17]. Variational inequalities have previously been utilized in general economic equilibrium only in the one-stage models in our papers [10, 11].

Along with providing support for existence arguments, variational inequalities have other properties worthy of attention from economists. A variational inequality problem constitutes a paradigm of analysis that can be compared with the classical paradigm of "n equations in n unknowns" but is able to cover vastly larger territory. An advanced theory is now available for studying perturbations and stability of their solutions [5]. Furthermore, they offer prospects of computational support which could lead to numerical experimentation with econometric models of equilibrium.

In order to proceed from a money-supported equilibrium, as defined above, to a variational inequality formulation, it is crucial to introduce auxiliary variables which complete the picture of "an n-dimensional condition in n unknowns." These variables are the Lagrange multipliers associated with the agents' budget constraints in optimality. Their values become part of the solution along with the prices and plans for market transactions; we speak then of a *enhanced* equilibrium.

The first step in arguing toward the existence of such an equilibrium is obviously then to make sure that Lagrange multipliers will be on hand. This follows from the concavity of utility and the assumption of ample survivability, which guarantees applicability of the Slater condition as a constraint qualification. The next step is to use the Lagrange multipliers in expressing necessary and sufficient conditions for optimality in an agent's maximization problem. The expression is achieved in the form of a *saddle point* condition on the associated Lagrangian function (which, by the way, would not be possible if utility were merely quasi-concave).

The saddle point condition can be translated in turn to a variational inequality of functional type. The market clearing condition in equilibrium can, on the other hand, be expressed through complementary slackness as a variational inequality of geometric type concerning the cones of normal vectors to a nonnegative orthant. These variational inequalities, as subconditions, can then be combined into a single composite variational inequality.

The composite variational inequality obtained in this way fully stands for an enhanced equilibrium, but it suffers from unboundedness which prevents immediate application of a criterion for the existence of a solution. An appeal to truncations must then be made, as is familiar in virtually all theory about the existence of economic equilibrium, even it if has to be carried out in a different way than in the past.

References

1. Arrow, K.J., Debreu, G.: Existence of an equilibrium for a competitive economy. Econometrica **22**, 265–290 (1954)
2. Arrow, K.J., Hahn, F.: Notes on sequence economies, transaction costs, and uncertainty. J. Econ. Theory **86**, 203–218 (1999)
3. Britz, V., Herings, P.J.-J., Predtetchinski, A.: Theory of the firm: bargaining and competitive equilibrium. (2009, preprint)
4. Cass, D.: Competitive equilibrium with incomplete financial markets. CARESS Working Paper Nl. 84-09, Univ. Pennsylvania 1984. Ultimately published, with corrections, in J. Math. Econ. **42**, 384–405 (2006)
5. Dontchev, A.D., Rockafellar, R.T.: Implicit Functions and Solution Mappings: A View From Variational Analysis. Monographs in Mathematics. Springer, Berlin (2009)
6. Facchinei, F., Pang, J.-S.: Finite-Dimensional Variational Inequalities and Complementarity Problems, vols. I and II. Springer Series in Operations Research. Springer, New York (2003)
7. Florig, M.: On irreducible economies. Ann. Econ. Stat. **61**, 184–199 (2001)
8. Florig, M.: Hierarchic competitive equilibria. J. Math. Econ. **35**, 515–546 (2001)
9. Grandmont, J.-M.: Temporary equilibrium. In: Working paper 2006–27. Institut National de la Statistique et des Études Économiques, Centre de Recherche en Économie et Statistique (2006)

10. Jofre, A., Rockafellar, R.T., Wets, R.J.-B.: A variational inequality model for determining an economic equilibrium of classical or extended type. In: Giannessi, F., Maugeri, A. (eds.) Variational Analysis and Applications, pp. 553–578. Springer, Berlin (2005)
11. Jofre, A., Rockafellar, R.T., Wets, R.J.-B.: Variational inequalities and economic equilibrium. Math. Oper. Res. **32**, 32–50 (2007)
12. Jofre, A., Rockafellar, R.T., Wets, R.J.-B.: General economic equilibrium with incomplete markets and money. Econometrica (2010, submitted)
13. Laitenberger, M.: Existence of financial markets equilibria with transaction costs. Ricerche Economiche **50**, 69–77 (1996)
14. Magill, M., Quinzii, M.: Real effects of money in general equilibrium. J. Math. Econ. **21**, 301–342 (1992)
15. Magill, M., Shafer, W.: Incomplete markets. In: Hildenbrand, W., Sonnenschein, H. (eds.) Chapter 30 in Handbook of Mathematical Economics, vol. IV. Elsevier Science, New York (1991)
16. Radner, R.: Existence of equilibrium of plans, prices, and price expectations. Econometrica **40**, 289–303 (1972)
17. Rockafellar, R.T.: Convex Analysis. Princeton University Press, Princeton, NJ (1970)
18. Rockafellar, R.T., Wets, R.J.-B. Variational Analysis. Grundlehren der Mathematischen Wissenschaften 317. Springer, Berlin (1997)
19. Skidelsky, R.: Return of the Master. BBS-Public Affairs, New York (2009)
20. Werner, J.: Equilibrium in economies with incomplete financial markets. J. Econ. Theory **36**, 110–119 (1985)

Adv. Math. Econ. 14, 197–218 (2011)

Advances in
MATHEMATICAL
ECONOMICS

Strong convergence theorems by hybrid methods for nonexpansive mappings with equilibrium problems in Banach spaces

Wataru Takahashi[1,2] **and Jen-Chih Yao**[2,*]

[1] Department of Mathematical and Computing Sciences,
Tokyo Institute of Technology, Japan
(e-mail: wataru@is.titech.ac.jp)

[2] Department of Applied Mathematics, National Sun Yat-sen University,
Kaohsiung 80424, Taiwan
(e-mail: yaojc@math.nsysu.edu.tw)

Received: January 11, 2010
Revised: July 9, 2010

JEL classification: C62, C68

Mathematics Subject Classification (2010): 47H05, 47H09, 47H20

Abstract. Our purpose in this paper is to prove strong convergence theorems by hybrid methods for nonexpansive mappings in a Banach space under appropriate conditions. We first prove a strong convergence theorem by the shrinking projection method for semi-positively homogeneous nonexpansive mappings with an equilibrium problem in a Banach space. Next, we obtain another strong convergence theorem by the monotone hybrid method for semi-positively homogeneous nonexpansive mappings with an equilibrium problem in a Banach space. These theorems are proved by using the concept of set convergence.

Key words: Nonexpansive mapping, fixed point, hybrid method, Mosco convergence, equilibrium problem, projection

* The research of the first author and the second author was partially supported by Grant-in-Aid for Scientific Research No. 19540167 from Japan Society for the Promotion of Science and by the grant NSC 98-2115-M-110-001, respectively.

DOI: 10.1007/978-4-431-53883-7_9,

1. Introduction

Let H be a real Hilbert space with inner product $\langle \cdot, \cdot \rangle$ and norm $\| \cdot \|$ and let C be a closed convex subset of H. Let T be a mapping of C into H. Then we denote by $F(T)$ the set of fixed points of T. A mapping $T : C \to H$ is called nonexpansive if $\|Tx - Ty\| \leq \|x - y\|$ for all $x, y \in C$.

In 2003, Nakajo and Takahashi [26] proved a strong convergence theorem for nonexpansive mappings in a Hilbert space by using the hybrid method [29] in mathematical programming: Let $T : C \to C$ be a nonexpansive mapping with $F(T) \neq \emptyset$ and let $\{\alpha_n\}$ be a real sequence in $[0, 1]$ such that $0 \leq \alpha_n \leq a < 1$ for all $n \in \mathbb{N}$. Define a sequence $\{x_n\}$ in C by $x_1 = x \in C$ and

$$\begin{cases} u_n = \alpha_n x_n + (1 - \alpha_n) T x_n, \\ C_n = \{z \in C : \|u_n - z\| \leq \|x_n - z\|\}, \\ Q_n = \{z \in C : \langle x_n - z, x - x_n \rangle \geq 0\}, \\ x_{n+1} = P_{C_n \cap Q_n} x, \quad \forall n \in \mathbb{N}, \end{cases}$$

where $P_{C_n \cap Q_n}$ is the metric projection of H onto $C_n \cap Q_n$. Then, $\{x_n\}$ converges strongly to an element z of $F(T)$, where $z = P_{F(T)} x$ and $P_{F(T)}$ is the metric projection of H onto $F(T)$. Recently, Takahashi, Takeuchi and Kubota [38] proposed a hybrid method called the shrinking projection method which is different from Nakajo and Takahashi method and then obtained a strong convergence theorem for nonexpansive mappings in a Hilbert space. Qin and Su [28] proposed also a hybrid method called the monotone hybrid method which is different from Nakajo and Takahashi method and Takahashi, Takeuchi and Kubota method. However, strong convergence theorems by such hybrid methods for nonexpansive mappings in a Hilbert space have not been extended to Banach spaces. Many authors have extended these convergence theorems to Banach spaces by using nonlinear mappings which are different from a nonexpansive mapping; see, for instance, [3, 16, 20–22, 27, 33, 37].

Let C be a nonempty closed convex subset of a Hilbert space H and let $f : C \times C \to \mathbb{R}$ be a bifunction. Then, the equilibrium problem is to find $z \in C$ such that

$$f(z, y) \geq 0, \quad \forall y \in C. \tag{1.1}$$

The set of such $z \in C$ is denoted by $EP(f)$, i.e.,

$$EP(f) = \{z \in C : f(z, y) \geq 0, \quad \forall y \in C\}.$$

Problem (1.1) is important in the sense that it includes, as special cases, optimization problems, variational inequalities, minimax problems, Nash equilibrium problem in noncooperative games and others; see, for instance, [6, 7, 24, 25, 30, 34, 35] for equilibrium problems in a Hilbert space.

Our purpose in this paper is to prove strong convergence theorems by hybrid methods for nonexpansive mappings in a Banach space under appropriate conditions. We first prove a strong convergence theorem by the shrinking projection method for semi-positively homogeneous nonexpansive mappings with an equilibrium problem in a Banach space. Next, we obtain another strong convergence theorem by the monotone hybrid method for semi-positively homogeneous nonexpansive mappings with an equilibrium problem in a Banach space. These theorems are proved by using the concept of set convergence.

2. Preliminaries

Let E be a real Banach space with norm $\|\cdot\|$ and let E^* be the dual of E. We denote the value of $y^* \in E^*$ at $x \in E$ by $\langle x, y^* \rangle$. When $\{x_n\}$ is a sequence in E, we denote the strong convergence of $\{x_n\}$ to $x \in E$ by $x_n \to x$ and the weak convergence by $x_n \rightharpoonup x$. The modulus δ of convexity of E is defined by

$$\delta(\epsilon) = \inf\left\{1 - \frac{\|x + y\|}{2} : \|x\| \leq 1, \|y\| \leq 1, \|x - y\| \geq \epsilon\right\}$$

for every ϵ with $0 \leq \epsilon \leq 2$. A Banach space E is said to be uniformly convex if $\delta(\epsilon) > 0$ for every $\epsilon > 0$. A uniformly convex Banach space is strictly convex and reflexive. Let C be a nonempty closed convex subset of a strictly convex and reflexive Banach space E. We know that for any $x \in E$, there exists a unique element $z \in C$ such that $\|x - z\| \leq \|x - y\|$ for all $y \in C$. Denote this element z by $P_C(x)$. Then P_C is called the metric projection of E onto C. The duality mapping J from E into 2^{E^*} is defined by

$$Jx = \{x^* \in E^* : \langle x, x^* \rangle = \|x\|^2 = \|x^*\|^2\}$$

for every $x \in E$. Let $U = \{x \in E : \|x\| = 1\}$. The norm of E is said to be Gâteaux differentiable if for each $x, y \in U$, the limit

$$\lim_{t \to 0} \frac{\|x + ty\| - \|x\|}{t} \tag{2.1}$$

exists. In this case, E is called smooth. We know that E is smooth if and only if J ia a single-valued mapping of E into E^*. We also know that E is reflexive if and only if J is surjective, and E is strictly convex if and only if J is one-to-one. Therefore, if E is a smooth, strictly convex and reflexive Banach space, then J is a single-valued bijection and in this case, the inverse mapping J^{-1} coincides with the duality mapping J_* on E^*. The norm of E is said to be uniformly Gâteaux differentiable if for each $y \in U$, the limit (2.1) is attained uniformly for $x \in U$. It is also said to be Fréchet differentiable if for each $x \in U$, the limit (2.1) is attained uniformly for $y \in U$. A Banach space E is called uniformly smooth if the limit (2.1) is attained uniformly for $x, y \in U$. It is known that if the norm of E is uniformly Gâteaux differentiable, then J is uniformly norm to weak* continuous on each bounded subset of E, and if the norm of E is Fréchet differentiable, then J is norm to norm continuous. If E is uniformly smooth, then J is uniformly norm to norm continuous on each bounded subset of E. For more details, see [31, 32].

We know the following result: Let E be a smooth, strictly convex and reflexive Banach space. Let C be a nonempty closed convex subset of E and let P_C be the metric projection of E onto C. Let $x_0 \in C$ and $x_1 \in E$. Then, $x_0 = P_C(x_1)$ if and only if

$$\langle x_0 - y, J(x_1 - x_0)\rangle \geq 0$$

for all $y \in C$, where J is the duality mapping of E.

Let C be a nonempty subset of E and let T be a mapping of C into E. We denote the set of all fixed points of T by $F(T)$. A mapping $T : C \to E$ is said to be nonexpansive if $\|Tx - Ty\| \leq \|x - y\|$ for all $x, y \in C$. A mapping $T : C \to E$ is quasi-nonexpansive if $F(T) \neq \emptyset$ and

$$\|Tx - y\| \leq \|x - y\| \tag{2.2}$$

for all $x \in C$ and $y \in F(T)$. If C is a closed convex subset of a strictly convex Banach space E and $T : C \to C$ is quasi-nonexpansive, then $F(T)$ is closed and convex; see Itoh and Takahashi [14]. If a nonexpansive mapping $T : E \to E$ is linear, then such a mapping is called contractive.

Let E be a reflexive, strictly convex and smooth Banach space. The function $\phi: E \times E \to (-\infty, \infty)$ is defined by

$$\phi(x, y) = \|x\|^2 - 2\langle x, Jy\rangle + \|y\|^2$$

for all $x, y \in E$, where J is the duality mapping of E; see [1, 15]. We have from the definition of ϕ that

$$\phi(x, y) = \phi(x, z) + \phi(z, y) + 2\langle x - z, Jz - Jy\rangle \tag{2.3}$$

for all $x, y, z \in E$. From $(\|x\|^2 - \|y\|^2) \leq \phi(x, y)$ for all $x, y \in E$, we can see that $\phi(x, y) \geq 0$. If E is additionally assumed to be strictly convex, then

$$\phi(x, y) = 0 \Longleftrightarrow x = y. \tag{2.4}$$

If C is a nonempty closed convex subset of a smooth, strictly and reflexive Banach space E, then for all $x \in E$ there exists a unique $z \in C$ (denoted by $\Pi_C x$) such that

$$\phi(z, x) = \min_{y \in C} \phi(y, x). \tag{2.5}$$

The mapping Π_C is called the generalized projection from E onto C; see Alber [1], Alber and Reich [2], and Kamimura and Takahashi [15]. The following theorem [15] is well known.

Theorem 2.1. *Let E be a uniformly convex and smooth Banach space and let $\{x_n\}$, $\{y_n\}$ be sequences in E such that $\{x_n\}$ or $\{y_n\}$ is bounded. If $\lim_{n\to\infty} \phi(x_n, y_n) = 0$, then $\lim_{n\to\infty} \|x_n - y_n\| = 0$.*

For a sequence $\{C_n\}$ of nonempty closed convex subsets of a reflexive Banach space E, define s-$\mathrm{Li}_n C_n$ and w-$\mathrm{Ls}_n C_n$ as follows: $x \in$s-$\mathrm{Li}_n C_n$ if and only if there exists $\{x_n\} \subset E$ such that $\{x_n\}$ converges strongly to x and $x_n \in C_n$ for all $n \in \mathbb{N}$. Similarly, $y \in$w-$\mathrm{Ls}_n C_n$ if and only if there exist a subsequence $\{C_{n_i}\}$ of $\{C_n\}$ and a sequence $\{y_i\} \subset E$ such that $\{y_i\}$ converges weakly to y and $y_i \in C_{n_i}$ for all $i \in \mathbb{N}$. If C_0 satisfies that

$$C_0 = \text{s-Li}_n C_n = \text{w-Ls}_n C_n, \tag{2.6}$$

we say that $\{C_n\}$ converges to C_0 in the sense of Mosco [23] and we write $C_0 =$M-$\lim_{n\to\infty} C_n$. It is easy to show that if $\{C_n\}$ is decreasing with respect to inclusion, then $\{C_n\}$ converges to $\bigcap_{n=1}^{\infty} C_n$ in the sense of Mosco. For more details, see [23]. We know the following theorem [13].

Theorem 2.2. *Let E be a smooth Banach space and let E^* have a Fréchet differentiable norm. Let $\{C_n\}$ be a sequence of nonempty closed convex subsets of E. If $C_0 =M$-$\lim_{n\to\infty} C_n$ exists and nonempty, then for each $x \in E$, $\Pi_{C_n} x$ converges strongly to $\Pi_{C_0} x$, where Π_{C_n} and Π_{C_0} are the generalized projections of E onto C_n and C_0, respectively.*

Let C be a nonempty closed subset of a smooth, strictly convex and reflexive Banach space E such that JC is closed and convex. For solving the equilibrium problem, let us assume that a bifunction $f : JC \times JC \to \mathbb{R}$ satisfies the following conditions:

(A1) $f(x^*, x^*) = 0, \quad \forall x^* \in JC$.
(A2) f is monotone, i.e., $f(x^*, y^*) + f(y^*, x^*) \le 0, \quad \forall x^*, y^* \in JC$.
(A3) $\lim_{t\downarrow 0} f(tz^* + (1-t)x^*, y^*) \le f(x^*, y^*), \quad \forall x^*, y^*, z^* \in JC$.
(A4) For each $x^* \in JC$, $y^* \mapsto f(x^*, y^*)$ is convex and lower semicontinuous.

Let E be a smooth Banach space E and let C be a nonempty subset of E. A mapping $T : C \to C$ is generalized nonexpansive [11] if $F(T) \neq \emptyset$ and

$$\phi(Tx, y) \le \phi(x, y) \tag{2.7}$$

for all $x \in C$ and $y \in F(T)$. For the fixed point problem of such mappings, see [8, 18, 19]. The following result is in Takahashi and Zembayashi [36]; see also [4–6].

Lemma 2.3. *Let C be a nonempty closed subset of a smooth, strictly convex and reflexive Banach space E such that JC is closed and convex. Let f be a bifunction from $JC \times JC$ into $\mathbb{R}$ satisfying (A1)–(A4). Then, for any $r > 0$ and $x \in E$, there exists a unique $z \in C$ such that*

$$f(Jz, Jy) + \frac{1}{r}\langle Jy - Jz, z - x\rangle \geq 0, \quad \forall y \in C.$$

Further, define $T_r x = \{z \in C : f(Jz, Jy) + \frac{1}{r}\langle Jy - Jz, z - x\rangle \geq 0, \forall y \in C\}$ for all $r > 0$ and $x \in E$ and $EP(f) = \{z \in C : f(Jz, Jy) \geq 0, \forall y \in C\}$. Then the following hold:

(1) T_r is single-valued.
(2) T_r is firmly generalized nonexpansive, i.e.,

$$\langle T_r x - T_r y, JT_r x - JT_r y\rangle \leq \langle x - y, JT_r x - JT_r y\rangle, \quad \forall x, y \in E.$$

(3) $F(T_r) = EP(f)$.
(4) $JEP(f)$ is closed and convex.
(5) $\phi(x, T_r x) + \phi(T_r x, q) \leq \phi(x, q), \quad \forall x \in E,\ q \in F(T_r)$.

Let E be a Banach space and let D be a nonempty closed subset of E. A mapping $R : E \to D$ is said to be sunny if

$$R(Rx + t(x - Rx)) = Rx, \quad \forall x \in E,\ \forall t \geq 0.$$

A mapping $R : E \to D$ is a retraction if $Rx = x$ for all $x \in D$. A nonempty subset D of a smooth Banach space E is said to be a generalized nonexpansive retract (resp. sunny generalized nonexpansive retract) of E if there exists a generalized nonexpansive retraction (resp. sunny generalized nonexpansive retraction) of E onto D. From [11], we know the following lemmas.

Lemma 2.4 (Ibaraki and Takahashi [11]**).** *Let E be a smooth, strictly convex and reflexive Banach space and let D be a nonempty closed subset of E. Then, a sunny generalized nonexpansive retraction of E onto D is uniquely determined.*

Lemma 2.5 (Ibaraki and Takahashi [11]**).** *Let E be a smooth, strictly convex and reflexive Banach space and let D be a nonempty closed subset of E. Suppose that there exists a sunny generalized nonexpansive retraction R of E onto D and let $(x, z) \in E \times D$. Then, the following hold:*

(1) $z = Rx$ if and only if $\langle x - z, Jy - Jz\rangle \leq 0, \ \forall y \in D$.
(2) $\phi(Rx, z) + \phi(x, Rx) \leq \phi(x, z)$.

In 2007, Kohsaka and Takahashi [17] proved the following results.

Lemma 2.6 (Kohsaka and Takahashi [17]). *Let E be a smooth, strictly convex and reflexive Banach space and let C_* be a nonempty closed convex subset of E^*. Suppose that Π_{C_*} is the generalized projection of E^* onto C_*. Then, R defined by $R = J^{-1}\Pi_{C_*}J$ is a sunny generalized nonexpansive retraction of E onto $J^{-1}C_*$.*

Lemma 2.7 (Kohsaka and Takahashi [17]). *Let E be a smooth, strictly convex and reflexive Banach space and let D be a nonempty subset of E. Then, the following conditions are equivalent*

(1) D is a sunny generalized nonexpansive retract of E.
(2) D is a generalized nonexpansive retract of E.
(3) JD is closed and convex.

In this case, D is closed.

Lemma 2.8 (Kohsaka and Takahashi [17]). *Let E be a smooth, strictly convex and reflexive Banach space and let D be a nonempty closed subset of E. Suppose that there exists a sunny generalized nonexpansive retraction R of E onto D and let $(x, z) \in E \times D$. Then, the following conditions are equivalent*

(1) $z = Rx$.
(2) $\phi(x, z) = \min_{y \in D} \phi(x, y)$.

From Ibaraki and Takahashi [12] we know the following lemma.

Lemma 2.9 (Ibaraki and Takahashi [12]). *Let E be a smooth, strictly convex and reflexive Banach space and let T be a generalized nonexpansive mapping of E into itself. Then, $F(T)$ is a sunny generalized nonexpansive retract of E.*

For more results concerning generalized nonexpansive mappings and sunny generalized nonexpansive retractions, see [9, 10].

3. Semi-positively homogeneous nonexpansive mappings

Let E be a Banach space and let C be a closed convex subset of E with $0 \in C$. Then, a mapping $T : C \to C$ is called semi-positively homogeneous near 0 if for any $x \in C$ with $x \neq 0$, there exists a real number t with $0 < t < 1$ such that $T(\alpha x) = \alpha T(x)$ for all $\alpha \in (0, t)$. In this section, we prove that a nonexpansive mapping $T : C \to C$ under an appropriate condition is generalized nonexpansive. Before proving it, we prove the following lemma.

Lemma 3.1. *Let E be a smooth Banach space and let C be a closed convex subset of E with $0 \in C$. Let $T : C \to C$ be a quasi-nonexpansive mapping which is semi-positively homogeneous near 0 and continuous at 0. Then, $T0 = 0$. Further, for any $x \in C$ and $m \in F(T)$,*

$$\langle x - Tx, Jm \rangle \leq 0.$$

Proof. First, we show $T0 = 0$. Fix $x \in C$ with $x \neq 0$. By the assumption of T, there exists a real number t with $0 < t < 1$ such that $T(\alpha x) = \alpha T(x)$ for all $\alpha \in (0, t)$. Since T is continuous at 0, we have that $T(\alpha x) \to T0$ as $\alpha \to 0$. On the other hand, we have that $\alpha Tx \to 0$ as $\alpha \to 0$. Then, we have $T0 = 0$.

Let $x \in C$ and $m \in F(T)$. Since T is quasi-nonexpansive, we have that

$$\begin{aligned}
\langle x - Tx, J(x - m) \rangle &= \langle x - Tx - (m - Tm), J(x - m) \rangle \\
&= \|x - m\|^2 - \langle Tx - Tm, J(x - m) \rangle \\
&\geq \|x - m\|^2 - \|Tx - Tm\| \|x - m\| \\
&\geq \|x - m\|^2 - \|x - m\|^2 \\
&= 0.
\end{aligned}$$

Let $x \in C$ with $x \neq 0$. Since there exists a real number t with $0 < t < 1$ such that $T(\alpha x) = \alpha T(x)$ for all $\alpha \in (0, t)$, we have

$$\langle \alpha x - \alpha Tx, J(\alpha x - m) \rangle = \langle \alpha x - T\alpha x, J(\alpha x - m) \rangle \geq 0$$

and hence

$$\langle x - Tx, J(\alpha x - m) \rangle \geq 0.$$

Letting $\alpha \to 0$, we have from the smoothness of E that

$$\langle x - Tx, J(-m) \rangle \geq 0$$

and hence $\langle x - Tx, J(m) \rangle \leq 0$. In the case of $x = 0$, it is obvious that $\langle x - Tx, J(m) \rangle \leq 0$. □

Theorem 3.2. *Let E be a smooth Banach space and let C be a closed convex subset of E with $0 \in C$. Let $T : C \to C$ be a quasi-nonexpansive mapping which is semi-positively homogeneous near* 0 *and continuous at* 0. *Then, $T : C \to C$ is a generalized nonexpansive mapping.*

Proof. Since $T0 = 0$, it is obvious that $F(T)$ is nonempty. Since T is quasi-nonexpansive, we have from $0 \in F(T)$ that for any $x \in C$,

$$\|Tx\| = \|Tx - 0\| \leq \|x - 0\| = \|x\|.$$

So, we have from Lemma 3.1 that for any $x \in C$ and $m \in F(T)$,

$$\begin{aligned}
\phi(Tx, m) &= \|Tx\|^2 - 2\langle Tx, Jm \rangle + \|m\|^2 \\
&\leq \|x\|^2 - 2\langle x, Jm \rangle + \|m\|^2 = \phi(x, m).
\end{aligned}$$

Therefore, T is a generalized nonexpansive mapping of C into itself. □

Let E be a Banach space and let C be a closed convex cone of E. Then, a mapping $T : C \to C$ is called positively homogeneous if $T(\alpha x) = \alpha T(x)$ for all $x \in C$ and $\alpha > 0$. Using Theorem 3.2, we obtain the following result.

Theorem 3.3. *Let E be a smooth Banach space and let C be a closed convex cone of E. Let $T : C \to C$ be a positively homogeneous nonexpansive mapping. Then, T is a generalized nonexpansive mapping.*

Proof. A closed convex cone C of E is a closed convex subset of E with $0 \in C$. A positively homogeneous nonexpansive mapping $T : C \to C$ is a quasi-nonexpansive mapping which is semi-positively homogeneous near 0 and continuous at 0. So, T is a generalized nonexpansive mapping of C into itself from Theorem 3.2. □

As a direct consequence of Theorem 3.3, we get the following result.

Theorem 3.4. *Let E be a smooth Banach space. Let $T : E \to E$ be a linear contractive mapping. Then, $T : E \to E$ is a generalized nonexpansive mapping.*

4. Strong convergence theorems

In this section, using Theorem 2.2 and Theorem 3.2, we prove a strong convergence theorem by the shrinking projection method [38] for linear contractive mappings with equilibrium problems in a Banach space.

Theorem 4.1. *Let E be a uniformly convex and uniformly smooth Banach space and let C be a closed convex subset of E with $0 \in C$ such that JC is closed and convex. Let $f : JC \times JC \to \mathbb{R}$ be a bifunction satisfying (A1)–(A4) and let S be a nonexpansive mapping which is semi-positively homogeneous near 0. Suppose $EP(f) \cap F(S) \neq \emptyset$, let $C_1 = C$ and let $\{x_n\} \subset C$ be a sequence generated by $x_1 = x \in C$ and*

$$\begin{cases} f(Jz_n, Jy) + \frac{1}{\lambda_n}\langle z_n - x_n, Jy - Jz_n\rangle \geq 0, & \forall y \in C, \\ y_n = \alpha_n x_n + (1-\alpha_n)Sz_n, \\ C_{n+1} = \{z \in C_n : \phi(y_n, z) \leq \phi(x_n, z)\}, \\ x_{n+1} = R_{C_{n+1}}x, \quad \forall n \in \mathbb{N}, \end{cases}$$

where $R_{C_{n+1}}$ is the sunny generalized nonexpansive retraction of E onto C_{n+1}, and $\{\alpha_n\} \subset [0, 1]$ and $\{\lambda_n\} \subset [0, \infty)$ are sequences such that

$$0 \leq \alpha_n \leq a < 1 \quad \text{and} \quad 0 < b \leq \lambda_n$$

for some $a, b \in \mathbb{R}$. Then, $\{x_n\}$ converges strongly to $z_0 = R_{F(S)\cap EP(f)}x$, where $R_{F(S)\cap EP(f)}$ is the sunny generalized nonexpansive retraction of E onto $F(S) \cap EP(f)$.

Proof. Put $z_n = T_{\lambda_n}x_n$ for each $n \in \mathbb{N}$ and take $z \in F(S) \cap EP(f)$. From $z = T_{\lambda_n}z$ and Lemma 2.3, we have that for any $n \in \mathbb{N}$,

$$\phi(z_n, z) = \phi(T_{\lambda_n}x_n, z) \le \phi(x_n, z). \tag{4.1}$$

We shall show that JC_n are closed and convex and $F(S) \cap EP(f) \subset C_n$ for all $n \in \mathbb{N}$. It is obvious from the assumption that $JC_1 = JC$ is closed and convex and $F(S) \cap EP(f) \subset C_1$. Suppose that JC_k is closed and convex, and $F(S)\cap EP(f) \subset C_k$ for some $k \in \mathbb{N}$. From the definition of ϕ, we know that for $z \in C_k$,

$$\begin{aligned} &\phi(y_k, z) \le \phi(x_k, z) \\ \iff &\|y_k\|^2 - \|x_k\|^2 - 2\langle y_k - x_k, Jz\rangle \le 0. \end{aligned}$$

So, JC_{k+1} is closed and convex. If $z \in F(S) \cap EP(f) \subset C_k$, then we have from (4.1) and Theorem 3.2 that

$$\begin{aligned} \phi(y_n, z) &= \phi(\alpha_n x_n + (1 - \alpha_n)Sz_n, z) \\ &\le \alpha_n\phi(x_n, z) + (1 - \alpha_n)\phi(z_n, z) \\ &\le \alpha_n\phi(x_n, z) + (1 - \alpha_n)\phi(x_n, z) \\ &= \phi(x_n, z). \end{aligned}$$

Hence, we have $z \in C_{k+1}$. By induction, we have that JC_n are closed and convex and $F(S) \cap EP(f) \subset C_n$ for all $n \in \mathbb{N}$. Since JC_n is closed and convex, from Lemma 2.7 there exists a unique sunny generalized nonexpansive retraction R_{C_n} of E onto C_n. We know from Lemma 2.6 that R_{C_n} is denoted by $J^{-1}\Pi_{JC_n}J$, where Π_{JC_n} is the generalized projection of E^* onto JC_n. Thus, $\{x_n\}$ is well-defined.

Since $\{JC_n\}$ is a decreasing sequence of nonempty closed convex subsets of E^* with respect to inclusion, it follows that

$$\emptyset \ne JF(S) \cap JEP(f) \subset \text{M-}\lim_{n\to\infty} JC_n = \bigcap_{n=1}^{\infty} JC_n. \tag{4.2}$$

Put $C_0^* = \bigcap_{n=1}^{\infty} JC_n$. Then, by Theorem 2.2 we have that $\{\Pi_{JC_{n+1}}Jx\}$ converges strongly to $x_0^* = \Pi_{C_0^*}Jx$. Since E^* has a Fréchet differentiable norm, J^{-1} is norm-to-norm continuous. So, we have

$$x_{n+1} = J^{-1}\Pi_{JC_{n+1}}Jx \to J^{-1}x_0^*.$$

To complete the proof, it is sufficient to show that $J^{-1}x_0^* = R_{F(S)\cap EP(f)}x$.

Since $x_n = R_{C_n}x$ and $x_{n+1} = R_{C_{n+1}}x \in C_{n+1} \subset C_n$, we have from Theorem 2.5 and (2.3) that

$$\begin{aligned} 0 &\leq 2\langle x - x_n, Jx_n - Jx_{n+1}\rangle \\ &= \phi(x, x_{n+1}) - \phi(x, x_n) - \phi(x_n, x_{n+1}) \\ &\leq \phi(x, x_{n+1}) - \phi(x, x_n). \end{aligned}$$

So, we get that

$$\phi(x, x_n) \leq \phi(x, x_{n+1}). \tag{4.3}$$

Further, since $x_n = R_{C_n}x$ and $z \in F(S) \cap EP(f) \subset C_n$, from Lemma 2.8 we have

$$\phi(x, x_n) \leq \phi(x, z). \tag{4.4}$$

So, we have that $\lim_{n\to\infty} \phi(x, x_n)$ exists. This implies that $\{x_n\}$ is bounded. Hence, $\{y_n\}$, $\{z_n\}$ and $\{Sz_n\}$ are also bounded. From Theorem 2.5, we have

$$\begin{aligned} \phi(x_n, x_{n+1}) &= \phi(R_{C_n}x, x_{n+1}) \\ &\leq \phi(x, x_{n+1}) - \phi(x, R_{C_n}x) \\ &= \phi(x, x_{n+1}) - \phi(x, x_n) \to 0. \end{aligned}$$

So, we have that

$$\phi(x_n, x_{n+1}) \to 0. \tag{4.5}$$

From $x_{n+1} \in C_{n+1}$, we also have that $\phi(y_n, x_{n+1}) \leq \phi(x_n, x_{n+1})$. So, we get that $\phi(y_n, x_{n+1}) \to 0$. Using Theorem 2.1, we have

$$\lim_{n\to\infty} \|y_n - x_{n+1}\| = \lim_{n\to\infty} \|x_n - x_{n+1}\| = 0.$$

So, we have

$$\|y_n - x_n\| \leq \|y_n - x_{n+1}\| + \|x_{n+1} - x_n\| \to 0. \tag{4.6}$$

Since $\|x_n - y_n\| = \|x_n - \alpha_n x_n - (1 - \alpha_n)Sz_n\| = (1 - \alpha_n)\|x_n - Sz_n\|$ and $0 \leq \alpha_n \leq a < 1$, we also have that

$$\|Sz_n - x_n\| \to 0. \tag{4.7}$$

Let $z \in F(S) \cap EP(f)$. Using $z_n = T_{\lambda_n}x_n$ and Lemma 2.3, we have that

$$\begin{aligned} \phi(x_n, z) &\geq \phi(x_n, T_{\lambda_n}x_n) + \phi(T_{\lambda_n}x_n, z) \\ &= \phi(x_n, z_n) + \phi(z_n, z) \end{aligned}$$

and hence

$$\phi(x_n, z_n) \leq \phi(x_n, z) - \phi(z_n, z).$$

From the definition of y_n and ϕ, we have $\phi(y_n, z) \leq \alpha_n \phi(x_n, z) + (1 - \alpha_n)\phi(z_n, z)$ and hence

$$\phi(z_n, z) \geq \frac{\phi(y_n, z) - \alpha_n \phi(x_n, z)}{1 - \alpha_n}.$$

Therefore, we have

$$\begin{aligned}\phi(x_n, z_n) &\leq \phi(x_n, z) - \frac{\phi(y_n, z) - \alpha_n \phi(x_n, z)}{1 - \alpha_n}\\ &= \frac{\phi(x_n, z) - \phi(y_n, z)}{1 - \alpha_n}.\end{aligned}$$

We also have

$$\begin{aligned}\phi(x_n, z) - \phi(y_n, z) &= \|x_n\|^2 - 2\langle x_n, Jz\rangle + \|z\|^2 - \|y_n\|^2\\ &\quad + 2\langle y_n, Jz\rangle - \|z\|^2\\ &= \|x_n\|^2 - \|y_n\|^2 - 2\langle x_n - y_n, Jz\rangle\\ &\leq |\|x_n\|^2 - \|y_n\|^2| + 2|\langle x_n - y_n, Jz\rangle|\\ &\leq \|x_n - y_n\|(\|x_n\| + \|y_n\|) + 2\|x_n - y_n\|\|Jz\|.\end{aligned}$$

Since $\lim_{n\to\infty} \|x_n - y_n\| \to 0$ from (4.6), we have

$$\lim_{n\to\infty} (\phi(x_n, z) - \phi(y_n, z)) = 0. \tag{4.8}$$

Since $0 \leq \alpha_n \leq a < 1$, we have $\lim_{n\to\infty} \phi(x_n, z_n) = 0$. So, from Theorem 2.1, we have

$$\|x_n - z_n\| \to 0. \tag{4.9}$$

From $y_n = \alpha_n x_n + (1 - \alpha_n)Sz_n$, we have $y_n - Sz_n = \alpha_n(x_n - Sz_n)$. So, from (4.7) we have

$$\|y_n - Sz_n\| = \alpha_n \|x_n - Sz_n\| \to 0. \tag{4.10}$$

Since

$$\begin{aligned}\|x_n - Sx_n\| &\leq \|x_n - y_n\| + \|y_n - Sz_n\| + \|Sz_n - Sx_n\|\\ &\leq \|x_n - y_n\| + \|y_n - Sz_n\| + \|z_n - x_n\|,\end{aligned}$$

from (4.6), (4.9) and (4.10) we have

$$\|x_n - Sx_n\| \to 0. \tag{4.11}$$

Since $x_n \to J^{-1}x_0^*$ and S is continuous, we have $J^{-1}x_0^* \in F(S)$. Next, let us show $J^{-1}x_0^* \in EP(f)$. From $x_n \to J^{-1}x_0^*$ and (4.9), we have $z_n \to J^{-1}x_0^*$. We have from $z_n = T_{\lambda_n}x_n$ that for any $y \in C$,

$$f(Jz_n, Jy) + \frac{1}{\lambda_n}\langle Jy - Jz_n, z_n - x_n\rangle \geq 0.$$

From (A2), we have

$$\frac{1}{\lambda_n}\langle Jy - Jz_n, z_n - x_n\rangle \geq f(Jy, Jz_n).$$

From $0 < b \leq \lambda_n$ and (4.9), we know

$$\lim_{n\to\infty} \frac{z_n - x_n}{\lambda_n} = 0.$$

So, we have

$$0 \geq f(Jy, x_0^*). \tag{4.12}$$

Put $z_t^* = tJy + (1-t)x_0^*$ for all $t \in (0, 1]$ and $y \in C$. Since JC is convex, we have $z_t^* \in JC$. From (A1), (A4) and (4.12), we have

$$\begin{aligned} 0 = f(z_t^*, z_t^*) &\leq tf(z_t^*, Jy) + (1-t)f(z_t^*, x_0^*) \\ &\leq tf(z_t^*, Jy) \end{aligned}$$

and hence

$$0 \leq f(z_t^*, Jy).$$

Letting $t \to 0$, we have from (A3) that for each $y \in C$,

$$0 \leq f(x_0^*, Jy). \tag{4.13}$$

This implies $J^{-1}x_0^* \in EP(f)$. So, we have that $J^{-1}x_0^* \in F(S) \cap EP(f)$. Put $z_0 = R_{F(S)\cap EP(f)}x$. Since $z_0 = R_{F(S)\cap EP(f)}x \subset C_{n+1}$ and $x_{n+1} = R_{C_{n+1}}x$, we have that

$$\phi(x, x_{n+1}) \leq \phi(x, z_0). \tag{4.14}$$

Hence, we have that

$$\begin{aligned}\phi(x, J^{-1}x_0^*) &= \|x\|^2 - 2\langle x, x_0^*\rangle + \|J^{-1}x_0^*\|^2 \\ &= \lim_{n\to\infty} \phi(x, x_n) \\ &\leq \phi(x, z_0).\end{aligned}$$

Therefore, we get $z_0 = J^{-1}x_0^*$. Hence, $\{x_n\}$ converges strongly to z_0. This completes the proof. □

As a direct consequence of Theorem 4.1, we have the following result.

Theorem 4.2. *Let E be a uniformly convex and uniformly smooth Banach space. Let $f : E^* \times E^* \to \mathbb{R}$ be a bifunction satisfying (A1)–(A4) and let S be a linear contractive mapping. Suppose $EP(f) \cap F(S) \neq \emptyset$, let $C_1 = E$ and let $\{x_n\} \subset E$ be a sequence generated by $x_1 = x \in E$ and*

$$\begin{cases} f(Jz_n, Jy) + \frac{1}{\lambda_n}\langle z_n - x_n, Jy - Jz_n\rangle \geq 0, \quad \forall y \in E, \\ y_n = \alpha_n x_n + (1-\alpha_n)Sz_n, \\ C_{n+1} = \{z \in C_n : \phi(y_n, z) \leq \phi(x_n, z)\}, \\ x_{n+1} = R_{C_{n+1}}x, \quad \forall n \in \mathbb{N}, \end{cases}$$

where $R_{C_{n+1}}$ is the sunny generalized nonexpansive retraction of E onto C_{n+1}, and $\{\alpha_n\} \subset [0, 1]$ and $\{\lambda_n\} \subset [0, \infty)$ are sequences such that

$$0 \leq \alpha_n \leq a < 1 \quad \text{and} \quad 0 < b \leq \lambda_n$$

for some $a, b \in \mathbb{R}$. Then, $\{x_n\}$ converges strongly to $z_0 = R_{F(S)\cap EP(f)}x$, where $R_{F(S)\cap EP(f)}$ is the sunny generalized nonexpansive retraction of E onto $F(S) \cap EP(f)$.

Proof. Since a linear contractive mapping S is a nonexpansive mapping which is semi-positively homogeneous near 0, we obtain the desired result by Theorem 4.1. □

We can make the coefficient condition $0 \leq \alpha_n \leq a < 1$ in Theorem 4.1 weaker. Next, we prove a strong convergence theorem by another hybrid method which was considered by [28].

Theorem 4.3. *Let E be a uniformly convex and uniformly smooth Banach space and let C be a closed convex subset of E with $0 \in C$ such that JC is closed and convex. Let $f : JC \times JC \to \mathbb{R}$ be a bifunction satisfying (A1)–(A4) and let S be a nonexpansive mapping which is semi-positively homogeneous near 0. Suppose $EP(f) \cap F(S) \neq \emptyset$, let $C_1 = Q_1 = C$ and let $\{x_n\} \subset C$ be a sequence generated by $x_1 = x \in C$ and*

$$\begin{cases} f(Jz_n, Jy) + \frac{1}{\lambda_n}\langle z_n - x_n, Jy - Jz_n\rangle \geq 0, & \forall y \in C, \\ y_n = \alpha_n x_n + (1-\alpha_n)Sz_n, \\ C_{n+1} = \{z \in C_n \cap Q_n : \phi(y_n, z) \leq \phi(x_n, z)\}, \\ Q_{n+1} = \{z \in C_n \cap Q_n : \langle Jx_n - Jz, x - x_n\rangle \geq 0\}, \\ x_{n+1} = R_{C_{n+1}\cap Q_{n+1}}x, & \forall n \in \mathbb{N}, \end{cases}$$

where $R_{C_{n+1}\cap Q_{n+1}}$ *is the sunny generalized nonexpansive retraction of E onto* $C_{n+1} \cap Q_{n+1}$, *and* $\{\alpha_n\} \subset [0, 1]$ *and* $\{\lambda_n\} \subset [0, \infty)$ *satisfy*

$$\liminf_{n\to\infty} \alpha_n < 1 \quad \text{and} \quad 0 < b \leq \lambda_n$$

for some $a, b \in \mathbb{R}$. *Then,* $\{x_n\}$ *converges strongly to* $z_0 = R_{F(S)\cap EP(f)}x$, *where* $R_{F(S)\cap EP(f)}$ *is the sunny generalized nonexpansive retraction of E onto* $F(S) \cap EP(f)$.

Proof. Put $z_n = T_{\lambda_n}x_n$ for each $n \in \mathbb{N}$ and take $z \in F(S) \cap EP(f)$. From $z = T_{\lambda_n}z$ and Lemma 2.3, we have that for any $n \in \mathbb{N}$,

$$\phi(z_n, z) = \phi(T_{\lambda_n}x_n, z) \leq \phi(x_n, z). \tag{4.15}$$

We shall show that JC_n and JQ_n are closed and convex and $F(S)\cap EP(f) \subset C_n \cap Q_n$ for all $n \in \mathbb{N}$. It is obvious that JQ_n is closed and convex. We show that JC_n is closed and convex. It is obvious that $JC_1 = JC$ is closed and convex. Suppose that JC_k is closed and convex for some $k \in \mathbb{N}$. From the definition of ϕ, we know that for $z \in C_k$,

$$\phi(y_n, z) \leq \phi(x_n, z)$$
$$\Longleftrightarrow \|y_n\|^2 - \|x_n\|^2 - 2\langle y_n - x_n, Jz\rangle \leq 0.$$

So, JC_{k+1} is closed and convex. By induction, we have that JC_n are closed and convex for all $n \in \mathbb{N}$. Next, we show $z \in F(S) \cap EP(f) \subset C_n \cap Q_n$ for all $n \in \mathbb{N}$. It is obvious that $F(S) \cap EP(f) \subset C_1 \cap Q_1$. Suppose that $F(S)\cap EP(f) \subset C_k \cap Q_k$ for some $k \in \mathbb{N}$. For $z \in F(S)\cap EP(f) \subset C_k \cap Q_k$, we have from (4.15) and Theorem 3.2 that

$$\begin{aligned} \phi(y_{k+1}, z) &= \phi(\alpha_{k+1}x_{k+1} + (1-\alpha_{k+1})Sz_{k+1}, z) \\ &\leq \alpha_{k+1}\phi(x_{k+1}, z) + (1-\alpha_{k+1})\phi(z_{k+1}, z) \\ &\leq \alpha_{k+1}\phi(x_{k+1}, z) + (1-\alpha_{k+1})\phi(x_{k+1}, z) \\ &= \phi(x_{k+1}, z). \end{aligned}$$

So, we have $z \in F(S)\cap EP(f) \subset C_{k+1}$. Further, since $x_k = R_{C_k\cap Q_k}x$, from Lemma 2.5 we have

$$\langle Jx_k - Jz, x - x_k\rangle \geq 0, \quad \forall z \in C_k \cap Q_k.$$

Since $F(S) \cap EP(f) \subset C_k \cap Q_k$, we have

$$\langle Jx_k - Jz, x - x_k \rangle \geq 0, \quad \forall z \in F(S) \cap EP(f).$$

This implies $F(S) \cap EP(f) \subset Q_{k+1}$. So, we have $F(S) \cap EP(f) \subset C_n \cap Q_n$ for all $n \in \mathbb{N}$. This means that $\{x_n\}$ is well-defined.

Since $\{JC_n \cap JQ_n\}$ is a decreasing sequence of nonempty closed convex subsets of E^* with respect to inclusion, it follows that

$$\emptyset \neq JF(S) \cap JEP(f) \subset \mathrm{M}-\lim_{n\to\infty} JC_n \cap JQ_n = \cap_{n=1}^{\infty}(JC_n \cap JQ_n). \tag{4.16}$$

Putting $C_0^* = \cap_{n=1}^{\infty}(JC_n \cap JQ_n)$, by Theorem 2.2 we have that $\{\Pi_{JC_{n+1}\cap JQ_{n+1}}Jx\}$ converges strongly to $x_0^* = \Pi_{C_0^*}Jx$. Since E^* has a Fréchet differential norm, J^{-1} is continuous. So, we have

$$x_{n+1} = J^{-1}\Pi_{JC_{n+1}\cap JQ_{n+1}}Jx \to J^{-1}x_0^*.$$

To complete the proof, it is sufficient to show that $J^{-1}x_0^* = R_{F(S)\cap EP(f)}x$.

Since $x_n = R_{Q_n}x$ and $x_{n+1} = R_{C_{n+1}\cap Q_{n+1}}x \subset Q_n$, we have

$$\begin{aligned} 0 &\leq 2\langle x - x_n, Jx_n - Jx_{n+1}\rangle \\ &= \phi(x, x_{n+1}) - \phi(x, x_n) - \phi(x_n, x_{n+1}) \\ &\leq \phi(x, x_{n+1}) - \phi(x, x_n). \end{aligned}$$

So, we get that

$$\phi(x, x_n) \leq \phi(x, x_{n+1}). \tag{4.17}$$

Further, since $x_n = R_{C_n\cap Q_n}x$ and $z \in F(S) \cap EP(f) \subset C_n$, we have

$$\phi(x, x_n) \leq \phi(x, z). \tag{4.18}$$

So, we have that $\lim_{n\to\infty}\phi(x, x_n)$ exists. This implies that $\{x_n\}$ is bounded. Hence, $\{y_n\}$, $\{z_n\}$ and $\{Sz_n\}$ are also bounded. From

$$\begin{aligned} \phi(x_n, x_{n+1}) &= \phi(R_{Q_n}x, x_{n+1}) \\ &\leq \phi(x, x_{n+1}) - \phi(x, R_{Q_n}x) \\ &= \phi(x, x_{n+1}) - \phi(x, x_n) \to 0, \end{aligned}$$

We have that

$$\phi(x_n, x_{n+1}) \to 0. \tag{4.19}$$

From $x_{n+1} \in C_{n+1}$, we have that $\phi(y_n, x_{n+1}) \leq \phi(x_n, x_{n+1})$. So, we get that $\phi(y_n, x_{n+1}) \to 0$. From Theorem 2.1, we have

$$\lim_{n\to\infty} \|y_n - x_{n+1}\| = \lim_{n\to\infty} \|x_n - x_{n+1}\| = 0.$$

So, we have

$$\|y_n - x_n\| \leq \|y_n - x_{n+1}\| + \|x_{n+1} - x_n\| \to 0. \tag{4.20}$$

From the assumption of $\{\alpha_n\}$, there exists a subsequence $\{\alpha_{n_i}\}$ of $\{\alpha_n\}$ converging to $\alpha_0 \in [0, 1)$. From $\|x_n - y_n\| = \|x_n - \alpha_n x_n - (1 - \alpha_n)Sz_n\| = (1 - \alpha_n)\|x_n - Sz_n\|$, we also have that

$$\|Sz_{n_i} - x_{n_i}\| \to 0. \tag{4.21}$$

Let $z \in F(S) \cap EP(f)$. Using $z_n = T_{\lambda_n} x_n$ and Lemma 2.3, we have that

$$\begin{aligned}\phi(x_n, z) &\geq \phi(x_n, T_{\lambda_n} x_n) + \phi(T_{\lambda_n} x_n, z)\\ &= \phi(x_n, z_n) + \phi(z_n, z)\end{aligned}$$

and hence

$$\phi(x_n, z_n) \leq \phi(x_n, z) - \phi(z_n, z).$$

From (4.15), we have $\phi(y_n, z) \leq \alpha_n \phi(x_n, z) + (1 - \alpha_n)\phi(z_n, z)$ and hence

$$\phi(z_n, z) \geq \frac{\phi(y_n, z) - \alpha_n \phi(x_n, z)}{1 - \alpha_n}.$$

Therefore, we have

$$\begin{aligned}\phi(x_n, z_n) &\leq \phi(x_n, z) - \frac{\phi(y_n, z) - \alpha_n \phi(x_n, z)}{1 - \alpha_n}\\ &= \frac{\phi(x_n, z) - \phi(y_n, z)}{1 - \alpha_n}.\end{aligned}$$

As in the proof of Theorem 4.1, we have

$$\begin{aligned}\phi(x_n, z) - \phi(y_n, z) &= \|x_n\|^2 - 2\langle x_n, Jz\rangle + \|z\|^2 - \|y_n\|^2 + 2\langle y_n, Jz\rangle - \|z\|^2\\ &\leq \|x_n - y_n\|(\|x_n\| + \|y_n\|) + 2\|x_n - y_n\|\|Jz\|.\end{aligned}$$

Since $\lim_{n\to\infty} \|x_n - y_n\| \to 0$, we have

$$\lim_{n\to\infty} (\phi(x_n, z) - \phi(y_n, z)) = 0. \tag{4.22}$$

So, we have $\lim_{n\to\infty} \phi(x_{n_i}, z_{n_i}) = 0$. From Theorem 2.1, we have

$$\|x_{n_i} - z_{n_i}\| \to 0. \tag{4.23}$$

Since $y_n = \alpha_n x_n + (1 - \alpha_n)Sz_n$, we have $y_n - Sz_n = \alpha_n(x_n - Sz_n)$. So, from (4.21) we have

$$\|y_{n_i} - Sz_{n_i}\| = \alpha_{n_i}\|x_{n_i} - Sz_{n_i}\| \to 0. \tag{4.24}$$

Since

$$\begin{aligned}\|x_{n_i} - Sx_{n_i}\| &\le \|x_{n_i} - y_{n_i}\| + \|y_{n_i} - Sz_{n_i}\| + \|Sz_{n_i} - Sx_{n_i}\| \\ &\le \|x_{n_i} - y_{n_i}\| + \|y_{n_i} - Sz_{n_i}\| + \|z_{n_i} - x_{n_i}\|,\end{aligned}$$

from (4.20), (4.23) and (4.24) we have

$$\|x_{n_i} - Sx_{n_i}\| \to 0. \tag{4.25}$$

Since $x_{n_i} \to J^{-1}x_0^*$ and S is continuous, we have $J^{-1}x_0^* \in F(S)$. Next, let us show $J^{-1}x_0^* \in EP(f)$. From (4.23), we have $z_{n_i} \to J^{-1}x_0^*$. Since $z_n = T_{\lambda_n}x_n$, we have that for any $y \in C$,

$$f(Jz_n, Jy) + \frac{1}{\lambda_n}\langle Jy - Jz_n, z_n - x_n\rangle \ge 0.$$

From (A2), we have

$$\frac{1}{\lambda_n}\langle Jy - Jz_n, z_n - x_n\rangle \ge f(Jy, Jz_n).$$

From $0 < b \le \lambda_n$, we have

$$\lim_{n\to\infty} \frac{z_{n_i} - x_{n_i}}{\lambda_{n_i}} = 0.$$

So, we have

$$0 \ge f(Jy, x_0^*). \tag{4.26}$$

Put $z_t^* = tJy + (1-t)x_0^*$ for all $t \in (0, 1]$ and $y \in C$. Since JC is convex, we have $z_t^* \in JC$. From (A1), (A4) and (4.26), we have

$$\begin{aligned}0 = f(z_t^*, z_t^*) &\le tf(z_t^*, Jy) + (1-t)f(z_t^*, x_0^*) \\ &\le tf(z_t^*, Jy)\end{aligned}$$

and hence

$$0 \le f(z_t^*, Jy).$$

Letting $t \to 0$, we have that for each $y \in C$,

$$0 \le f(x_0^*, Jy). \tag{4.27}$$

This implies $J^{-1}x_0^* \in EP(f)$. So, we have that $J^{-1}x_0^* \in F(S)\cap EP(f)$. Put $z_0 = R_{F(S)\cap EP(f)}x$. Since $z_0 = R_{F(S)\cap EP(f)}x \subset C_{n+1}$ and $x_{n+1} = R_{C_{n+1}}x$, we have that

$$\phi(x, x_{n+1}) \le \phi(x, z_0). \tag{4.28}$$

So, we have that

$$\begin{aligned}\phi(x, J^{-1}x_0^*) &= \lim_{n\to\infty} \phi(x, x_n) \\ &\le \phi(x, z_0).\end{aligned}$$

Therefore, we get $z_0 = J^{-1}x_0^*$. Hence, $\{x_n\}$ converges strongly to z_0. This completes the proof. □

As a direct consequence of Theorem 4.3, we have the following result.

Theorem 4.4. *Let E be a uniformly convex and uniformly smooth Banach space Let $f : E^* \times E^* \to \mathbb{R}$ be a bifunction satisfying (A1)–(A4) and let S be a linear contractive mapping. Suppose $EP(f) \cap F(S) \neq \emptyset$, let $C_1 = Q_1 = E$ and let $\{x_n\} \subset E$ be a sequence generated by $x_1 = x \in E$ and*

$$\begin{cases} f(Jz_n, Jy) + \frac{1}{\lambda_n}\langle z_n - x_n, Jy - Jz_n\rangle \ge 0, \quad \forall y \in E, \\ y_n = \alpha_n x_n + (1-\alpha_n)Sz_n, \\ C_{n+1} = \{z \in C_n \cap Q_n : \phi(y_n, z) \le \phi(x_n, z)\}, \\ Q_{n+1} = \{z \in C_n \cap Q_n : \langle Jx_n - Jz, x - x_n\rangle \ge 0\}, \\ x_{n+1} = R_{C_{n+1}\cap Q_{n+1}}x, \quad \forall n \in \mathbb{N}, \end{cases}$$

where $R_{C_{n+1}\cap Q_{n+1}}$ is the sunny generalized nonexpansive retraction of E onto $C_{n+1} \cap Q_{n+1}$, and $\{\alpha_n\} \subset [0, 1]$ and $\{\lambda_n\} \subset [0, \infty)$ satisfy

$$\liminf_{n\to\infty} \alpha_n < 1 \quad \text{and} \quad 0 < b \le \lambda_n$$

for some $a, b \in \mathbb{R}$. Then, $\{x_n\}$ converges strongly to $z_0 = R_{F(S)\cap EP(f)}x$, where $R_{F(S)\cap EP(f)}$ is the sunny generalized nonexpansive retraction of E onto $F(S) \cap EP(f)$.

Proof. Since a linear contractive mapping S is a nonexpansive mapping which is semi-positively homogeneous near 0, we obtain the desired result by Theorem 4.3. □

References

1. Alber, Y.I.: Metric and generalized projections in Banach spaces: properties and applications. In: Kartsatos, A.G. (ed.) Theory and Applications of Nonlinear Operators of Accretive and Monotone Type, pp. 15–50. Marcel Dekker, New York (1996)

2. Alber, Y.I., Reich, S.: An iterative method for soling a class of nonlinear operator equations in Banach spaces. PanAm. Math. J. **4**, 39–54 (1994)
3. Aoyama, K., Takahashi, W.: Strong convergence theorems for a family of relatively nonexpansive mappings in Banach spaces. Fixed Point Theory **8**, 143–160 (2007)
4. Aoyama, K., Kimura, Y., Takahashi, W.: Maximal monotone operators and maximal monotone functions for equilibrium problems. J. Convex Anal. **15**, 395–409 (2008)
5. Aoyama, K., Kohsaka, F., Takahashi, W.: Three generalizations of firmly nonexpansive mappings: their relations and continuity properties. J. Nonlinear Convex Anal. **10**, 131–147 (2009)
6. Blum, E., Oettli, W.: From optimization and variational inequalities to equilibrium problems. Math. Student **63**, 123–145 (1994)
7. Combettes, P.L., Hirstoaga, A.: Equilibrium programming in Hilbert spaces. J. Nonlinear Convex Anal. **6**, 117–136 (2005)
8. Dhompongsa, S., Fupinwong, W., Takahashi, W., Yao, J.-C.: Fixed point theorems for nonlinear mappings and strict convexity of Banach spaces. J. Nonlinear Convex Anal. **11**, 175–183 (2010)
9. Honda, T., Takahashi, W.: Nonlinear projections and generalized conditional expectations in Banach spaces. Taiwanese J. Math. (to appear)
10. Honda, T., Ibaraki, T., Takahashi, W.: Duality theorems and convergence theorems for nonlineaqr mappings in Banach spaces. Int. J. Math. Stat. **6**, 46–64 (2010)
11. Ibaraki, T., Takahashi, W.: A new projection and convergence theorems for the projections in Banach spaces. J. Approx. Theory **149**, 1–14 (2007)
12. Ibaraki, T., Takahashi, W.: Generalized nonexpansive mappings and a proximal-type algorithm in Banach spaces. Contemp. Math. (to appear)
13. Ibaraki, T., Kimura, Y., Takahashi, W.: Convergence theorems for generalized projections and maximal monotone operators in Banach spaces. Abst. Appl. Anal. **2003**, 621–629 (2003)
14. Itoh, S., Takahashi, W.: The common fixed point theory of singlevalued mappings and multivalued mappings. Pac. J. Math. **79**, 493–508 (1978)
15. Kamimura, S., Takahashi, W.: Strong convergence of a proximal-type algorithm in a Banach apace. SIAM J. Optim. **13**, 938–945 (2002)
16. Kimura, Y., Takahashi, W.: On a hybrid method for a family of relatively nonexpansive mappings in a Banach apace. J. Math. Anal. Appl. **357**, 356–363 (2009)
17. Kohsaka, F., Takahashi, W.: Generalized nonexpansive retractions and a proximal-type algorithm in Banach spaces. J. Nonlinear Convex Anal. **8**, 197–209 (2007)

18. Kohsaka, F., Takahashi, W.: Existence and approximation of fixed points of firmly nonexpansive-type mappings in Banach spaces. SIAM J. Optim. **19**, 824–835 (2008)
19. Kohsaka, F., Takahashi, W.: Fixed point theorems for a class of nonlinear mappings related to maximal monotone operators in Banach spaces. Arch. Math. **91**, 166–177 (2008)
20. Matsushita, S., Takahashi, W.: Weak and strong convergence theorems for relatively nonexpansive mappings in Banach spaces. Fixed Point Theory Appl. **2004**, 37–47 (2004)
21. Matsushita, S., Takahashi, W.: A strong convergence theorem for relatively nonexpansive mappings in a Banach space. J. Approx. Theory **134**, 257–266 (2005)
22. Matsushita, S., Takahashi, W.: Approximating fixed points of nonexpansive mappings in a Banach space by metric projections. Appl. Math. Comput. **196**, 422–425 (2008)
23. Mosco,U.: Convergence of convex sets and of solutions of variational inequalities. Adv. Math. **3**, 510–585 (1969)
24. Moudafi, A.: Weak convergence theorems for nonexpansive mappings and equilibrium problems. J. Nonlinear Convex Anal. **9**, 37–43 (2008)
25. Moudafi, A., Théra, M.: Proximal and dynamical approaches to equilibrium problems. Lecture Notes in Economics and Mathematical Systems, vol. 477, pp. 187–201. Springer, Berlin (1999)
26. Nakajo, K., Takahashi, W.: Strong convergence theorems for nonexpansive mappings and nonexpansive semigroups. J. Math. Anal. Appl. **279**, 372–379 (2003)
27. Ohsawa, S., Takahashi, W.: Strong convergence theorems for resolvents of maximal monotone operator. Arch. Math. **81**, 439–445 (2003)
28. Qin, X., Su, Y.: Strong convergence of monotone hybrid method for fixed point iteration processes. J. Syst. Sci. Complex. **21**, 474–482 (2008)
29. Solodov, M.V., Svaiter, B.F.: Forcing strong convergence of proximal point iterations in a Hilbert space. Math. Program. **87**, 189–202 (2000)
30. Tada, A., Takahashi, W.: Strong convergence theorem for an equilibrium problem and a nonexpansive mapping. J. Optim. Theory Appl. **133**, 359–370 (2007)
31. Takahashi, W.: Convex Analysis and Approximation of Fixed Points (Japanese). Yokohama Publishers, Yokohama (2000)
32. Takahashi, W.: Nonlinear Functional Analysis. Yokohama Publishers, Yokohama (2000)
33. Takahashi, W.: Proximal point algorithms and four resolvents of nonlinear operators of monotone type in Banach spaces. Taiwanese J. Math. **12**, 1883–1910 (2008)

34. Takahashi, S., Takahashi, W.: Viscosity approximation methods for equilibrium problems and fixed point problems in Hilbert spaces. J. Math. Anal. Appl. **331**, 506–515 (2007)
35. Takahashi, S., Takahashi, W.: Strong convergence theorems for a generalized equilibrium problem anda nonexpansive mapping in a Hilbert space. Nonlinear Anal. **69**, 1025–1033 (2008)
36. Takahashi, W., Zembayashi, K.: A strong convergence theorem for the equilibrium problem with a bifunction defined on the dual space of a Banach space. In: Dhompongsa, S., Goebel, K., Kirk, W.A., Plubtieng, S., Sims, B., Suantai, S. (eds.) Fixed Point Theory and Its Applications, pp. 197–209. Yokohama Publishers, Yokohama (2008)
37. Takahashi, W., Zembayashi, K.: Strong and weak convergence theorem for equilibrium problems and relatively nonexpansive mappings in Banach spaces. Nonlinear Anal. **70**, 45–57 (2009)
38. Takahashi, W., Takeuchi, Y., Kubota, R.: Strong convergence theorems by hybrid methods for families of nonexpansive mappings in Hilbert spaces. J. Math. Anal. Appl. **341**, 276–286 (2008)

Programme

Workshop on Mathematical Economics, 2009

Date: November 13(Fri.) – 15(Sun.), 2009
Venue: Lecture Hall, East Research Building,
Keio University
2-15-45 Mita, Minato-ku, Tokyo 108-8345, JAPAN

organized by
Research Center for
Mathematical Economics
cosponsored by
Oak Society

information
Toru Maruyama
Department of Economics
Keio University
2-15-45 Mita, Minato-ku, Tokyo
TEL: 03-3453-4511 ex. 23271
FAX: 03-5427-1578
e-mail: maruyama@econ.keio.ac.jp

November 13 (Friday) * : Speaker

Morning

Chair: Ryo Nagata (Waseda University)

9:00–10:00 **Norihisa Sato** (Waseda University)
Satiation and Existence of Competitive Equilibrium

10:00–11:00 **Keita Owari** (Hitotsubashi University)
Robust Utility Maximization with Random Endowment and Indifference Valuation

11:10–12:10 **Hisatoshi Tanaka** (Waseda University)
The Bracketing Number of Single Index Functions

Afternoon

Chair: Hidetoshi Komiya (Keio University)

13:30–14:30 **P. Jean-Jacques Herings*** (Maastricht University), A. Predtetchinski
One-dimensional Bargaining with Markov Recognition Probabilities

14:30–15:30 **Wataru Takahashi** (National Sun Yat-sen University)
Equilibrium Problems and Nonlinear Operators in Economic Theory

Chair: Koichiro Takaoka (Hitotsubashi University)

16:00–17:00 **Takuji Arai** (Keio University)
Convex Risk Measures on Orlicz Spaces — Convolution and Shortfall —

17:00–18:00 **Koji Kuroda*** (Nihon University), J. Masukawa, J. Murai
Stock Price Process and Long Memory in Trade Signs

Reception

18:30–20:00 Tsunamachi Mitsui Club

November 14 (Saturday)

Morning

Chair: Seiichi Iwamoto (Kyushu University)

9:00–10:00 **Shin Sato** (Fukuoka University)
Adjacent Manipulation: A New Criterion on Manipulability of Social Choice Functions

10:00–11:00 **Boris Mordukhovich** (Wayne St. University)
Set-Valued Optimization in Welfare Economics

11:10–12:10 **Tyrrell Rockafellar*** (University of Washington), R. Wets, A. Jofre
A Time-Embedded Real-Asset Framework for General Economic Equilibrium

Afternoon

Chair: Hideyuki Adachi (Onomichi University)

13:30–14:30 **Ali Khan*** (Johns Hopkins University), A. Piazza
Classical Turnpike Theory and the Economics of Forestry: The Discounted Case

14:30–15:30 **Takashi Kamihigashi** (Kobe University)
Rational Asset Price Bubbles in Small Open Economies

Chair: Shinichi Suda (Keio University)

16:00–17:00 **Kazuya Kamiya** (University of Tokyo)
Hysteresis in Dynamic General Equilibrium Models with Cash-in-Advance Constraints

17:00–18:00 **Nobusumi Sagara** (Hosei University)
A Lyapunov-type Theorem for Nonadditive Vector Measures

November 15 (Sunday)

Morning

Chair: Hiroshi Matano (University of Tokyo)

9:00–10:00 **Chiaki Hara** (Kyoto University)
Effectively Complete Asset Markets "tentative"

10:00–11:00 **Louis Nirenberg** (New York University)
A Geometic Problem and the Hopf Lemma

11:10–12:10 **Alexander Ioffe** (Israel Institute of Technology)
Variational Analysis and Nonsmooth Regular Economics

Afternoon

Chair: Akihiko Takahashi (University of Tokyo)

13:30–14:30 **Yumiharu Nakano** (Tokyo Institute of Technology)
Quantile Hedging for Defaultable Claims

14:30–15:30 **Shigeo Kusuoka** (University of Tokyo)
Approximation of Expectation of Diffusion Processes

Chair: Kunitake Ito (Kyoto University)

16:00–17:30 **S. Todd Lowry** (Washington & Lee University)
Pythagorean Mathematical Idealism and the Framing of Economic and Political Theory

Seiich Iwamoto, the director

Tyrell Rockafellar

Ali Khan, Boris Mordukhovich

P. Jean-Jacques Herings, Chiaki Hara

Alexander Ioffe, Rozalia Ioffe, Yoko Kawamata

Jeanne Marie Aubin, Louis Nirenberg

Subject Index

Instructions for Authors

A. General

1. Papers submitted for publication will be considered only if they have not been and will not be published elsewhere without permission from the publisher and the Research Center for Mathematical Economics.

2. Every submitted paper will be subject to review. The names of reviewers will not be disclosed to the authors or to anybody not involved in the editorial process.

3. The authors are asked to transfer the copyright to their articles to Springer if and when these are accepted for publication.

The copyright covers the exclusive and unlimited rights to reproduce and distribute the article in any form of reproduction. It also covers translation rights for all languages and countries.

4. Manuscript must be written in English. Its pdf file should be submitted by e-mail: maruyama@econ.keio.ac.jp

Office of Advances in
Mathematical Economics
c/o Professor Toru Maruyama
Department of Economics
Keio University
2-15-45, Mita Minato-ku,
Tokyo 108-8345, JAPAN

5. Twenty-five (25) **offprints** of each paper will be supplied free of charge. Additional offprints can be ordered at cost price.

B. Preparation of Manuscript

1. Manuscripts should be submitted in the pdf format. If this is not possible, two printouts of the manuscript must be submitted to the above postal address.

Manuscripts should be written in **LaTeX**. Please use Springer's LaTeX macro package (download from `ftp://ftp.springer.de/pub/tex/latex/svjour/global/`). After acceptance, sending the original source (including all style files and figures) and a pdf (compiled output) are required.

Authors wishing to include figures, tables, or text passages that have already been published elsewhere are required to **obtain permission** from the copyright owner(s) for both the print and online format.

2. **The title page** should include:
- The name(s) of the author(s)
- A concise and informative title
- The affiliation(s) and address(es) of the author(s)
- The e-mail address, telephone and fax numbers of the corresponding author

Please provide an **abstract** less than 100 words. The abstract should not contain any undefined abbreviations or unspecified references.

Please provide 4 to 6 **keywords** which can be used for indexing purposes.

3. Please use the decimal system of **headings** with no more than three levels.

Abbreviations should be defined at first mention and used consistently thereafter.

Footnotes can be used to give additional information, which may include the citation of a reference included in the reference list. They should not consist solely of a reference citation, and they should never include the bibliographic details of a reference. They should also not contain any figures or tables. Footnotes to the text are numbered consecutively; those to tables should be indicated by superscript lower-case letters (or asterisks for significance values and other statistical data). Footnotes to the title or the authors of the article are not given reference symbols. Always use footnotes instead of endnotes.

4. **The Journal of Economic Literature index number (JEL classification)** should be indicated and the statement of the **2000 Mathematics Subject Classification (MSC) numbers** is desirable. You can check JEL classification with Internet at `http://ideas.repec.org/JEL/` as well as 2000 MSC numbers at `http://www.ams.org/msc`.

5. **Main text**: All **tables and figures** must be cited in the text and numbered consecutively with Arabic numerals according to the sequence in which they are cited.

For each table, please supply a table caption (title) explaining the components of the table. Identify any previously published material by giving the original source in the form of a reference at the end of the table caption. When preparing your tables and figures, size them to fit in the column width.

Short **equations** can be run in with the text. Equations that are displayed on a separate line should be numbered.

6. **Reference citations** in the text should be identified by numbers in square brackets. Some examples:
1. Negotiation research spans many disciplines [3].
2. This result was later contradicted by Becker and Seligman [5].
3. This effect has been widely studied [1-3, 7].

The list of references should only include works that are cited in the text and that have been published or accepted for publication. Personal communications and unpublished works should only be mentioned in the text. Do not use footnotes or endnotes as a substitute for a reference list. The entries in the list should be numbered consecutively.

•*Journal article*
Hamburger, C.: Quasimonotonicity, regularity and duality for nonlinear systems of partial differential equations. Ann. Mat. Pura. Appl. 169, 321- 354 (1995)
•*Article by DOI*
Slifka, M.K., Whitton, J.L. Clinical implications of dysregulated cytokine production. J Mol Med. (2000) doi:10.1007/s001090000086
•*Book*
Geddes, K.O., Czapor, S.R., Labahn, G.: Algorithms for Computer Algebra. Kluwer, Boston (1992)
•*Book chapter*
Broy, M.: Software engineering - from auxiliary to key technologies. In: Broy, M., Denert, E. (eds.) Software Pioneers, pp. 10-13. Springer, Heidelberg (2002)
•*Online document*
Cartwright, J.: Big stars have weather too. IOP Publishing PhysicsWeb. `http://physicsweb.org/articles/news/11/6/16/1` (2007). Accessed 26 June 2007

Please use the standard abbreviation of a journal's name according to the ISSN List of Title Word Abbreviations, see `http://www.issn.org/2-22660-LTWA.php`

7. The purpose of the **author proof** is to check for typesetting or conversion errors and the completeness and accuracy of the text, tables and figures. Substantial changes in content, e.g., new results, corrected values, title and authorship, are not allowed without the approval of the Editor. After online publication, further changes can only be made in the form of an Erratum, which will be hyperlinked to the article.

8. Please use the standard **mathematical notation** for formulae, symbols etc.: Italic for single letters that denote mathematical constants, variables, and unknown quantities Roman/upright for numerals, operators, and punctuation, and commonly defined functions or abbreviations, e.g., cos, det, e or exp, lim, log, max, min, sin, tan, d (for derivative) Bold for vectors, tensors, and matrices.

Kolibri 360 GmbH
Ferdinand-Jühlke-Straße 7
99095 Erfurt, Deutschland
produktsicherheit@kolibri360.de